AF781381

A Path to the Next Generation of U.S. Banknotes

Keeping Them Real

Committee on Technologies to Deter Currency Counterfeiting

Board on Manufacturing and Engineering Design

Division on Engineering and Physical Sciences

NATIONAL RESEARCH COUNCIL
OF THE NATIONAL ACADEMIES

THE NATIONAL ACADEMIES PRESS
Washington, D.C.
www.nap.edu

THE NATIONAL ACADEMIES PRESS 500 Fifth Street, N.W. Washington, DC 20001

NOTICE: The project that is the subject of this report was approved by the Governing Board of the National Research Council, whose members are drawn from the councils of the National Academy of Sciences, the National Academy of Engineering, and the Institute of Medicine. The members of the committee responsible for the report were chosen for their special competences and with regard for appropriate balance.

This study was supported by Contract No. TEP-05-0002 between the National Academy of Sciences and the U.S. Department of the Treasury, Bureau of Engraving and Printing. Any opinions, findings, conclusions, or recommendations expressed in this publication are those of the authors and do not necessarily reflect the views of the organizations or agencies that provided support for the project.

International Standard Book Number-13: 978-0-309-10578-1
International Standard Book Number-10: 0-309-10578-1

Cover: Design by Steven Coleman. Images courtesy of the U.S. Department of Engraving and Printing. Reprinted with permission.

Available in limited quantities from:

Board on Manufacturing and Engineering Design
500 Fifth Street, N.W.
Washington, DC 20001
bmed@nas.edu
http://www.nationalacademies.edu/bmed

Additional copies of this report are available from the National Academies Press, 500 Fifth Street, N.W., Lockbox 285, Washington, DC 20055; (800) 624-6242 or (202) 334-3313 (in the Washington metropolitan area); Internet, http://www.nap.edu.

Copyright 2007 by the National Academy of Sciences. All rights reserved.

Printed in the United States of America

THE NATIONAL ACADEMIES

Advisers to the Nation on Science, Engineering, and Medicine

The **National Academy of Sciences** is a private, nonprofit, self-perpetuating society of distinguished scholars engaged in scientific and engineering research, dedicated to the furtherance of science and technology and to their use for the general welfare. Upon the authority of the charter granted to it by the Congress in 1863, the Academy has a mandate that requires it to advise the federal government on scientific and technical matters. Dr. Ralph J. Cicerone is president of the National Academy of Sciences.

The **National Academy of Engineering** was established in 1964, under the charter of the National Academy of Sciences, as a parallel organization of outstanding engineers. It is autonomous in its administration and in the selection of its members, sharing with the National Academy of Sciences the responsibility for advising the federal government. The National Academy of Engineering also sponsors engineering programs aimed at meeting national needs, encourages education and research, and recognizes the superior achievements of engineers. Dr. Charles M. Vest is president of the National Academy of Engineering.

The **Institute of Medicine** was established in 1970 by the National Academy of Sciences to secure the services of eminent members of appropriate professions in the examination of policy matters pertaining to the health of the public. The Institute acts under the responsibility given to the National Academy of Sciences by its congressional charter to be an adviser to the federal government and, upon its own initiative, to identify issues of medical care, research, and education. Dr. Harvey V. Fineberg is president of the Institute of Medicine.

The **National Research Council** was organized by the National Academy of Sciences in 1916 to associate the broad community of science and technology with the Academy's purposes of furthering knowledge and advising the federal government. Functioning in accordance with general policies determined by the Academy, the Council has become the principal operating agency of both the National Academy of Sciences and the National Academy of Engineering in providing services to the government, the public, and the scientific and engineering communities. The Council is administered jointly by both Academies and the Institute of Medicine. Dr. Ralph J. Cicerone and Dr. Charles M. Vest are chair and vice chair, respectively, of the National Research Council.

www.national-academies.org

COMMITTEE ON TECHNOLOGIES TO DETER CURRENCY COUNTERFEITING

ROBERT E. SCHAFRIK, GE Aviation, *Chair*
MARTIN A. CRIMP, Michigan State University
CHARLES B. DUKE, University of Rochester
ALAN H. GOLDSTEIN, Alfred University
ELIZABETH A. HOLM, Sandia National Laboratories
PRADEEP K. KHOSLA, Carnegie Mellon University
CAROLYN R. MERCER, NASA Glenn Research Center
STEPHEN M. POLLOCK, University of Michigan
ARTHUR J. RAGAUSKAS, Georgia Institute of Technology
JOHN A. ROGERS, University of Illinois at Urbana-Champaign
BARTON RUBENSTEIN, Rubenstein Studios
MICHAEL A. SMITH, France Telecom R&D
GARY K. STARKWEATHER, Microsoft (retired)
DENNIS J. TREVOR, OFS Laboratories

Staff

MICHAEL MOLONEY, Study Director (from February 2006)
TONI MARÉCHAUX, Study Director (until February 2006)
MARTA VORNBROCK, Research Associate (until July 2006)
TERI THOROWGOOD, Administrative Coordinator
LAURA TOTH, Senior Project Assistant (until June 2006)

BOARD ON MANUFACTURING AND ENGINEERING DESIGN

H.M. REININGA, Rockwell Collins, Inc., *Chair*
THOMAS HARTWICK, Adviser, Snohomish, Washington, *Vice Chair*
RON BLACKWELL, AFL-CIO
PAUL B. GERMERAAD, Intellectual Assets, Inc.
MANISH MEHTA, National Center for Manufacturing Sciences
NABIL Z. NASR, Rochester Institute of Technology
DENISE F. SWINK, Adviser, Germantown, Maryland
ALFONSO VELOSA III, Gartner, Inc.
BEVLEE A. WATFORD, Virginia Polytechnic Institute and State University
JOHN F. WHITE, Altarum

Staff

GARY FISCHMAN, Director

Preface

This report serves three purposes. First, at the request of the U.S. Department of the Treasury's Bureau of Engraving and Printing (BEP), the Committee on Technologies to Deter Currency Counterfeiting undertook a systematic investigation of the trends in digital imaging and printing technologies in order to assess evolving counterfeiting threats to U.S. currency. This task was accomplished through a review of current literature and practice under the guidance of members of the committee who are experts on these topics.

Second, the committee generated ideas for potential new features of Federal Reserve notes (FRNs) that could provide effective countermeasures to these threats. These ideas were subsequently evaluated using a two-step process. The first step consisted of highlighting features that the committee evaluated to be difficult for at least one class of counterfeiter to simulate while also being easy for the public and/or experienced cash handlers to detect; for some of these features, authentication requires assistance from a simple device. Features that survived this first round were then evaluated for key development risks and issues, which include the durability of the feature, its aesthetic appearance as part of a U.S. banknote, social acceptability issues (for example, toxicity of the materials, privacy concerns, and so on), and the estimated production cost of the feature. For those innovative features that have not been used in a security-feature application, the committee highlighted particular durability, cost, and other relevant concerns as top risk areas that should be evaluated early in the development program. Depending on the maturity level of a feature, it was considered as implementable in the intermediate term (less than 7 years) or the long term (more than 7 years).

The primary sources of data used by the committee in conducting these analyses were the expert knowledge of the committee members themselves, who are knowledgeable practitioners of most of the technologies applied by these new features. Consultation with the BEP, the U.S. Secret Service, and BEP vendor and advisory personnel also characterized this phase of the committee's activities. Care has been taken to ensure that the material cited in the report is available from open, public sources.

Finally, the third purpose served by this report relates to the committee's realization, toward the conclusion of its deliberations, that it had generated so many new feature concepts that a structured development process would be necessary to determine the feasibility of the most promising features. Thus, the committee provided an overview of a requirements-driven development process that is commonly employed by the aerospace industry. This type of process could be adapted to develop an advanced-generation currency.

The committee's report thus constitutes a comprehensive assessment of current technology-enabled counterfeiting threats, an enumeration of potentially effective countermeasures, and an indication of how the BEP might implement these countermeasures in a cost-effective and timely fashion. This report necessarily does not contain an exhaustive list of all possible new features. Additional technologies undoubtedly could be successfully employed to improve counterfeit deterrence. However, the committee believes that the process followed in conducting this study would be applicable to the analysis and development of other candidate features. It is anticipated that the audience for this report will extend beyond the sponsors at the BEP, to include government decision makers not intimately familiar with banknote security features, those performing research that could be relevant to advanced security features, interested members of the public, and so on. Therefore, the report provides context and background information so that its discussion is understandable to a variety of readers who are not necessarily banknote-security experts.

The title of this report, *A Path to the Next Generation of U.S. Banknotes: Keeping Them Real*, is a reflection of the challenge that the Department of the Treasury, through the BEP, faces in maintaining the worldwide security of U.S. banknotes without compromising their unique character and recognizability.

Three themes that are elaborated on in the body of the report emerged from the committee's discussions and analysis:

- First, digital printing technology is continuing to advance at a rapid pace, driven in large measure by the overwhelming success of digital photography. Therefore, within 10 years, low-skill amateurs will be able to duplicate almost any two-dimensional image. In a sense, the "battle of the printed image" will be lost to the counterfeiters. This advancing threat can best be met with new features that will allow the BEP to stay a step or two ahead of the counterfeiters. Owing to banknote design and production lead times, a "step or two ahead" translates to staying more than 5 years ahead of the most advanced technology that will be available to counterfeiters. Intrinsic limits in printing resolution, registration, pattern layouts, and inks associated with these printing methods motivate a search for radically different materials and manufacturing concepts.

- Second, a paradigm shift in security features seems possible. For instance, there appears to be tremendous potential to make fundamental modifications to the banknote substrate—that is, the paper in current notes. New feature design concepts are possible, such as incorporating heterogeneous materials and active elements into the substrate. There is an opportunity to leverage the large national effort in nano- and biotechnology to provide unique, highly secure features. These new approaches employ materials and technologies that have not necessarily been proven in other banknote applications.
- Third, in order to be realized, these advances in security features require a proactive approach to their development. The current materials used in Federal Reserve notes are well known and proven, with a wealth of experience gained over many years. Incorporating into security features new materials that possess expanded functional performance is not a step that can be taken without considerable additional technical development and cost-to-benefit justification. Moreover, while resistance to counterfeiting is vital, another important aspect of banknotes is that they are a manufactured product used daily by millions of people around the world. Therefore, for each of the billions of notes produced each year, features must be reproduced reliably. Notes must also be durable during normal use and able to survive folding, crumpling, and occasional laundering. Their design should be aesthetically pleasing. Finally, as with any manufactured product, banknote production should be cost-effective. A successful research and development effort must address each of these multifaceted objectives. Development must also span the entire sequence, from demonstrating the concept of a new counterfeit-resistant feature to its design incorporation and production in finished banknotes. The committee outlines a banknote development process that meets these criteria.

Historically, the BEP has been quite attuned to the threat of counterfeiting. Indeed, the bureau was established during the Civil War as a key element in the national strategy to reduce the volume of counterfeit currency flooding the Union. In recent years, the bureau has recognized that modern information technology could lead to entirely new types of counterfeiting threats. Consequently, over the past two decades, it has asked the National Research Council (NRC) to perform several studies to assess and characterize these evolving threats.[1] The BEP has further initiated a series of currency design changes aimed at reducing the vulnerability of U.S. banknotes to counterfeiting.

Recognizing the evolving threat of counterfeiting created by digital imaging and reprographics technologies, the BEP requested that the NRC undertake a new study to provide the bureau with up-to-date information on the factors that would allow it to produce designs to enhance the security of U.S. Federal Reserve notes. The committee's statement of task is presented in Appendix A, along with a description of the organization of the report. The NRC appointed the Committee on Technologies to Deter Currency Counterfeiting to carry out this statement of task. The committee members have expertise encompassing a broad range of disciplines and fields, knowledge of which was necessary in order to assess the future of the digital reprographic threat and to evaluate new technologies that could provide a paradigm shift for counterfeit-deterrence features. Appendix F presents biographical sketches of the committee members. Members of the committee are knowledgeable in analog and digital imaging and printing, art, biomaterials, computer software and hardware engineering, decision analysis and operations research, materials science and engineering, nanomaterials, optics, optical materials, paper science, and systems engineering. By design, the committee had no experts in currency design and production or in the various aspects of counterfeiting. Invited speakers and site visits supplemented the committee's knowledge base, especially in these areas.

The committee met a total of six times.[2] The meetings in May and July 2005 and in March and June 2006 included sessions that were open to the public; the remaining meetings were devoted to the preparation of the report.

Complementing the committee meetings, a number of site visits were conducted so that members could gain an appreciation of banknote production, counterfeit detection, and verification equipment. The committee thanks the following organizations and companies for making their personnel, facilities, and time available:

- Bureau of Engraving and Printing, Washington, D.C.;
- U.S. Secret Service, Washington, D.C.;

[1]The following reports, published by the National Academy Press, Washington D.C., have been issued by the NRC in response to these requests: *Advanced Reprographic Systems: Counterfeiting Threat Assessment and Deterrent Measures* (1985), *Counterfeit Threats and Deterrent Measures* (1987), and *Counterfeit Deterrent Features for the Next-Generation Currency Design* (1993).

[2]May 23-24, 2005, and July 21-22, 2005, in Washington, D.C.; October 10-11, 2005, in Woods Hole, Massachusetts; March 23-24, 2006, and June 1-2, 2006, in Washington, D.C.; and September 14-15, 2006, in Irvine, California.

- Crane and Company, Dalton, Massachusetts; and
- Cummins-Allison Corporation, Mount Prospect, Illinois.

The committee is grateful to the following individuals, who presented invited briefings on specific areas relevant to the management, design, and production of banknotes; security features; and machine verification of banknotes: Sara Church, Bank of Canada; Peter Crean, Xerox Corporation; Mark Crickett, De La Rue; John Haslop, De La Rue; Annette Jaffe, Jaffe Consulting; James Jonza, 3M Corporation; Ely Sachs, Massachusetts Institute of Technology; Jeff Thom, California Council of the Blind; and Stuart Thompson, Note Printing Australia.

The interim report of the study[3] assessed the counterfeiting threats to FRNs resulting from new technology, as specified in Task 1 of the committee's charge. This final report applies the assessment of these threats in evaluating new banknote features; the interim report is integrated into this report.

During the course of this study, the committee's deliberations also highlighted the perhaps greater threat of counterfeiting that may be perpetrated through cybercrimes relating to electronic funds transfer and digital currency. Considering these threats is clearly beyond the scope of this study, but the committee suggests that this important subject should be examined in the future to identify ways to further enhance U.S. economic security.

The committee is grateful to the following U.S. government personnel who took the time to share their perspectives with the committee: Lenore Clarke, Lisa DiNunzio, Larry Felix, Thomas A. Ferguson, Goutam Gupta, Kalyan Maitra, and Robert Stone of the Bureau of Engraving and Printing; Eugenie Foster from the Federal Reserve Board; and Douglas Albright and Lorelei Pagano of the U.S. Secret Service.

Finally, the committee acknowledges the support provided by the staff members of the National Research Council, including Laura Toth, Marta Vornbrock, Teri Thorowgood, Toni Maréchaux, Gary Fischman, and Michael Moloney.

Robert E. Schafrik, *Chair*
Committee on Technologies to Deter
Currency Counterfeiting

[3]National Research Council. 2006. Is That Real?: Identification and Assessment of the Counterfeiting Threat for U.S. Banknotes. Washington, D.C.: The National Academies Press.

Acknowledgment of Reviewers

This report has been reviewed in draft form by individuals chosen for their diverse perspectives and technical expertise, in accordance with procedures approved by the National Research Council's Report Review Committee. The purpose of this independent review is to provide candid and critical comments that will assist the institution in making its published report as sound as possible and to ensure that the report meets institutional standards for objectivity, evidence, and responsiveness to the study charge. The review comments and draft manuscript remain confidential to protect the integrity of the deliberative process. We wish to thank the following individuals for their review of this report:

Norbert S. Baer, New York University,
David R. Clarke, University of California, Santa Barbara,
Amy Crook, Not Dead Yet Studios,
Thomas Elder, U.S. Department of Agriculture,
Mitchell J. Feigenbaum, Rockefeller University,
Thomas A. Ferguson, U.S. Department of the Treasury (retired),
Thomas S. Hartwick, Independent Consultant,
George W. Lynch, Hewlett-Packard,
Alan G. Miller, The Boeing Company, and
Johannes Schaede, KBA-Giori SA.

Although the reviewers listed above have provided many constructive comments and suggestions, they were not asked to endorse the conclusions or recommendations, nor did they see the final draft of the report before its release. The review of this report was overseen by Pierre C. Hohenberg, New York University. Appointed by the National Research Council, he was responsible for making certain that an independent examination of this report was carried out in accordance with institutional procedures and that all review comments were carefully considered. Responsibility for the final content of this report rests entirely with the authoring committee and the institution.

Contents

Summary

At the request of the U.S. Department of the Treasury's Bureau of Engraving and Printing (BEP), the Committee on Technologies to Deter Currency Counterfeiting investigated trends in digital imaging and printing technologies to assess evolving counterfeiting threats to U.S. currency. The most significant threat is represented by nonprofessional, independent individuals with limited, if any, traditional counterfeiting skills. These opportunist, casual counterfeiters pose a particularly onerous risk to the integrity of currency, given the diffuse nature of their counterfeiting activities, which are enabled by affordable, highly capable, advanced reprographic and digital technology. This technology will continue to improve, primarily in response to digital photographic market demand for high-quality reproductions. Within 5 to 10 years, advances in image-processing tools and digital printing technology will allow a casual counterfeiter to attain the level of skill of a computer graphics art specialist. The committee also highlighted a newer threat—the use of the Internet to share best practices and sourcing of specialty materials, permitting the quality of the counterfeits to improve rapidly.

The BEP has adopted a successful proactive strategy of adding features to U.S. banknotes to enhance counterfeit deterrence. The counterfeiting rate of U.S. banknotes is one of the lowest of any major currency. The grand challenge is to uphold this record while maintaining the "look and feel" of traditional U.S. banknotes. Security features make the currency readily recognizable. Several strategies have been pursued to make the counterfeiter's job more challenging:

- The addition of features to the substrate (the paper in current U.S. notes), exploiting its three-dimensional character—for example, an embedded security strip;
- The use of an image that is visible in transmitted light but which cannot be directly imaged in reflected light by a copier or scanner—for example, a watermark;
- The employment of difficult-to-reproduce printed features—for example, color-shifting ink and microprint; and
- An improvement in the recognizability of printed images—for example, the use of an enlarged, off-center portrait.

This report identifies new feature ideas that exploit 21st-century materials and technologies to deter 21st-century counterfeiting of banknotes. An appropriate mix of these feature ideas can form the basis for extending the proactive strategy a step or two ahead of the technology available to counterfeiters. This strategy requires that government efforts be directed at feature innovation rather than being focused solely on the integration of features developed by others.

The committee generated ideas and devised and applied an evaluation process for a variety of potential new features that could provide effective countermeasures to these threats. In general, these features employ technologies that are not being developed for currency applications, although several that are in use in other currencies are suggested, since they can be more quickly implemented to provide an added measure of deterrence for a period of time.

In order to refine the list of potential features, the committee found it helpful to describe the counterfeiting threat by means of a basic analytical model that maps the flow of counterfeit currency from its production to its removal from circulation. The committee recognized that the two most important parameters for assessing the effectiveness of a feature are (1) how well it deters simulation and (2) how well it aids in authentication. These parameters formed the basis for determining which features should be selected for further consideration. This report necessarily does not contain an exhaustive list of all possible new features. Additional technologies undoubtedly could be successfully employed to improve counterfeit deterrence. However, the committee believes that the process followed in conducting this study would be applicable to the analysis and development of other candidate features.

On the basis of the maturity of the various technologies for security feature applications, the committee divided the feature ideas into two categories—(1) those that could be fully developed and ready for insertion in a banknote within 7 years and (2) those that would likely require longer than 7 years. Those feature ideas for the longer term were revolutionary and were considered "feature platforms," since many different feature options could be generated from these technologies. The game-changing technologies were derived from considering what could be possible in terms of applying advanced technology to security features. They require special materials and/or special manufacturing technologies, such as microfabrication.

The committee anticipates that the battle for the two-dimensional reflected-light image will be lost to opportunist counterfeiters owing to advances in reprographic technology. The features that require a longer development time do not depend on reflected light. Thus, advances in currency features will progress from reliance on printed optical features to reliance on features based on advanced materials and processes that would be far above the simulation capability of opportunist counterfeiters.

The feature ideas that survived the committee's evaluation process are listed below and described in detail in the report. Some of them can serve as upgrades and/or replacements for current features, while others are quite revolutionary. These features can be combined in different ways to provide a layered defense against a wide array of counterfeiters.

- The addition of features to the substrate exploiting its three-dimensional character—examples are anomalous currency space, digitally encrypted substrate, engineered cotton fibers, fiber-infused substrate, Fresnel lens for microprinting self-authentication, and window.
- The use of a transmitted-light image that cannot be directly imaged by a copier or scanner—an example is a see-through registration feature.
- The employment of printed features that are difficult for electronic printers to duplicate—examples are high-complexity spatial patterns, and nanoprint.
- The exploitation of the high pressure of intaglio printing[1] to effect surface characteristics—an example is grazing-incidence optical patterns.
- The addition of high-technology optical devices that produce dramatic visual effects—examples are hybrid diffractive optical variable devices, refractive microoptic arrays, and subwavelength optical elements.
- The local alteration of the tactile feel of the banknote, passively or actively—examples are chemical sensors, e-substrate, NiTi shape memory and superelastic responsive materials, tactilely active electronic features, tactile variant substrate, and thermoresponsive optically variable devices.

[1]The Bureau of Engraving and Printing prints currency on high-speed, sheet-fed rotary presses that are capable of printing over 8,000 sheets per hour. Printing plates are covered with ink, and then the surface of each plate is wiped clean, allowing the ink to remain in the design and letter grooves of the plates. Each sheet is then forced, under extremely heavy pressure (estimated at 20 tons), into the finely recessed lines of the printing plate to pick up the ink. The printing impression is three-dimensional in effect and requires the combined handiwork of highly skilled artists, steel engravers, and plate printers. The surface of the note feels slightly raised, while the reverse side feels slightly indented. This process is called intaglio printing.

- Sophisticated printed images that employ special inks and dyes that provide a controlled absorption, scattering, and/or fluorescent signature—examples are color image saturation, metameric ink patterns, and nanocrystal pigments.

As counterfeiting technology continues to improve, the committee foresees an increased interest by cash handlers in using simple generic devices (such as a pen-light) to aid in authenticating banknotes. This trend could also capture the interest of the public so that simple, readily accessible devices could be commonly used to validate banknotes. Many of the innovative security features would be readily detected by a simple device or machine. Features that might be useful to the blind are so noted in the committee's feature evaluations as discussed in the report.

The feature ideas described in the report require a dedicated research and development (R&D) program to establish their feasibility for application in banknotes and to make them ready for full-rate production implementation. The committee suggests that a significant annual budget would provide the seed funding to initiate a balanced R&D program consisting of low-risk and higher-risk, game-changing technologies. This would allow the BEP to implement a very proactive counterfeit-deterrence strategy. The BEP currently has no such R&D program.

The committee generated so many new feature concepts that a structured development process would be necessary—for example, adapting an existing technology- and product-development process to the specific task of currency design and production. Development priority would take many factors into account, such as the estimated deterrent effectiveness against a projected threat, the need for a balanced mix of features to provide a layered defense against the range of threats, ensuring a high degree of recognizability by the different types of cash handlers, and the estimated time line and implementation cost. These considerations would be matched against the BEP's internal assessment of the most critical security-feature needs and metrics.

1

Background and Motivation for the Study

This report is about identifying potential new features that will make use of 21st-century materials and technologies to deter 21st-century counterfeiting of banknotes. Deterring counterfeiting is important because the use of money to facilitate financial transactions is an essential element of the world's economies. Ensuring the value and stability of the nation's currency is critical to the maintenance of a healthy economy. Protecting confidence in the genuineness of the physical representation of that currency is an important aspect of that effort.

This chapter provides a discussion of some of the background and motivation for this study.[1] It presents an introduction to the current state of counterfeiting and introduces an analysis of counterfeiting that informs the generation of new ideas for the next generations of U.S. banknotes. The committee found that developing a good understanding of these issues—see Box 1-1 on the committee's work process—was an essential foundation (1) to its efforts to identify technologies, both existing and emerging, that pose the most significant counterfeiting threats to Federal Reserve notes (FRNs) (ideas that are developed further in Chapters 2 and 3) and (2) to its efforts to identify features, materials, and technologies to deter counterfeiting of FRNs (discussed in Chapters 4 and 5 and in Appendixes C and D). Chapter 6 discusses a proactive strategy for further developing new features for implementation.

---

[1]See Appendix A for the study's statement of task and a summary of the organization of this report, with a mapping onto the various elements of the task.

BOX 1-1
Committee Work Process: Identifying Novel Counterfeit-Deterrence Features

As part of this study's process of identifying and developing ideas for security features of Federal Reserve notes (FRNs), and in order to better understand the deterrence and authentication value of features used in current banknotes, the committee first carried out the following activities:

- Collected background material on the counterfeiting problem: its extent and location, types of counterfeiters and their methods, skill sets, and technologies;
- Gathered data and made judgments on how current features deter simulation by each class of counterfeiter and for each technology; and
- Forecasted technology trends in order to predict upcoming feature vulnerabilities.

The committee then conducted the following tasks:

- Collected information on features, their use, and simulation;
- Examined current banknote features in some detail, both for U.S. and international currencies;
- Surveyed commercially available features used currently in currency and high-security documents and reviewed studies on their use and effectiveness (the sources used are referenced in footnotes throughout the report);
- Examined actual counterfeit notes, both for U.S. currency (during U.S. Secret Service site visits) and currency from other countries;[1]
- Identified new feature concepts that span a wide range of technological readiness, from (intermediate-term) low-technology to (long-term) pie-in-the-sky;[2]
- Applied criteria to refine the list;[3] and
- Further explored the most promising ideas.[4]

---

[1]Inspecting U.S. banknote counterfeits provided the committee with important insights into which features are easily simulated and which are not. Because the counterfeit notes that the committee saw were in most cases passed and then ultimately detected, their makeup indicated which features are apparently important and which are generally ignored during note authentication. Observing passed counterfeits reinforced the committee's sense of the need to design notes with counterfeit deterrence in mind. Inspecting passed non-U.S. counterfeits was helpful in understanding the effectiveness of a larger feature set than is present on FRNs.

[2]In particular, the long-term feature ideas were generated by committee brainstorming.

[3]The features chosen for inclusion in the report emerged from an evaluation scheme developed by the committee that rated the motivation for adopting each feature—why it is difficult to simulate and how recognizable it is expected to be.

[4]The features were subsequently grouped into intermediate term (anticipated to have less than a 7-year development time before incorporation into a banknote) and long term (longer than a 7-year development time).

THE EVOLUTION OF MONEY

The physical representation of money has evolved over the centuries. The transformation from paper currency to electronic forms of money is clearly the latest chapter in this saga. The value of the transactions handled by electronic financial networks dwarfs the cash economies of the world.[2] Thus, the counterfeiting of electronic forms of money and its use for criminal purposes is an urgent and strategic issue for modern law enforcement. However, it is not the focus of this report, which addresses the counterfeiting of physical currency, specifically, FRNs.[3]

The modern world is still far from being a cashless society. Regardless of their lack of intrinsic value, banknotes continue to be ubiquitous and useful for a number of reasons:[4]

- *Access.* Many people do not have credit cards or checking accounts and only use cash transactions.
- *Anonymity.* Cash purchases preserve privacy and anonymity.
- *Convenience.* Cash enables rapid, low-technology sales transactions, which can take place without access to machine authentication or even electricity.
- *Acceptability.* Banknotes can serve as a universally accepted medium of exchange. The U.S. banknote is well accepted throughout the world because people understand that it is backed by confidence in U.S. economic power. Also, as a stable currency, it provides a measure of protection against inflation for people in countries with unstable currencies.

At the beginning of the 21st century, in the same way that "$" is a global symbol for "money," U.S. currency is considered to be a worldwide symbol of security and integrity. The unique combination of design, paper and inks, and printing technology makes a U.S. banknote unique and one of the most recognized symbols worldwide. Maintaining this symbol and what it stands for is among the duties of the U.S. Department of the Treasury's Bureau of Engraving and Printing (BEP).

In the early history of the United States, the design of paper currency changed many times. Since 1928, however, changes in the design have preserved the overall "architecture" of U.S. currency; that is, all U.S. currency in circulation has the same size and feel, and the historical figures and national symbols remain unchanged for each denomination. Appendix B describes the features on current U.S. notes,

[2]National Research Council. 2005. Network Science. Washington, D.C.: The National Academies Press.

[3]B. Grow. 2006. Gold rush. Business Week 3966:68-69.

[4]Ultimately, banknote designers take these "requirements" into account when selecting the portfolio of features for a banknote.

instituted in part to deter the creation of counterfeit notes. It is significant that while many changes have been instituted, there has never been a recall or demonetization of U.S. currency already in circulation.

Despite being one of the most respected items in this culture, a banknote can also be one of the most mistreated. No other item of such value is routinely folded, crumpled, soiled, laundered, and otherwise ill-treated throughout its useful life. To survive such challenges, U.S. banknotes meet daunting physical requirements. They are manufactured reliably in large quantities and are durable over time. In addition, features are placed on banknotes to allow authentication, indicate their denomination, and deter counterfeiting.[5] Recent issues of U.S. currency have included specific counterfeit-deterrent features to deter the threats from the advances in reprographic technologies mentioned in the following sections and discussed in more detail in Chapter 3.

The current $1 note was issued in 1929, and its design has remained essentially unchanged since then. The $2 note in use today was introduced in 1976, and its design has remained unchanged as well. The $5, $10, $20, $50, and $100 bills were redesigned beginning in 1990 to include several new security features that vary depending on the denomination. Most notable among the features added to FRNs are the introduction of offset printing (of colors other than green and black), color-shifting ink, watermarks, and colored threads. These features have been added while the characteristic, highly recognizable appearance of U.S. currency has been retained.

THE EVOLUTION OF COUNTERFEITING

Counterfeiting has existed almost as long as there has been money.[6] Though counterfeiting is predominantly a criminal activity, it has also been used by a number of countries as a weapon of war. Nations have a strong obligation to protect the integrity of their banknotes against attempts to make illegal copies. The deterrence of counterfeiting is an important element of government that is required to maintain confidence in a nation's currency both domestically and internationally. Because any original banknote can be duplicated—provided the materials, equip-

[5]The addition of new features begins with the New Currency Design Task Force, which has representatives from the U.S. Department of the Treasury, the Federal Reserve System, the U.S. Secret Service, and the BEP. The Task Force makes its recommendations to the Advanced Counterfeit Deterrence Steering Committee, also composed of representatives of the Treasury Department, Federal Reserve, Secret Service, and BEP. The Steering Committee then makes recommendations for the new design and security features to the Secretary of the Treasury, who has the statutory authority to approve such changes.

[6]Counterfeiting of currency has existed in the United States from the birth of the country. For some brief notes on its history, see <http://www.secretservice.gov/counterfeit.shtml>. Accessed February 2007.

ment, and expertise are accessible—responsible states endeavor to use materials and techniques that are not generally available and that present as many obstacles as possible to would-be counterfeiters. In the United States, the BEP and the U.S. Secret Service[7] were established to combat pervasive counterfeiting during the Civil War. At that time, it was estimated that one-third to one-half of the currency in circulation was counterfeit.

For hundreds of years, counterfeiting had required considerable artistic and technical skill, as well as substantial resources. Until recently, the primary counterfeiting threat arose from organized professional criminals and, in a few instances, from hostile states. These types of counterfeiters were relatively large enterprises that presented multiple opportunities for tracking by law enforcement. Because the production of realistic copies of a sophisticated banknote was quite an expensive proposition, the quality of the counterfeits oftentimes was not high, enabling the public to spot forgeries readily.

This scenario began to change during the 1980s with the advent of advanced reprographic systems and the accessibility of highly capable and inexpensive graphics software tools running on readily available workstations and desktop computers. Counterfeiting of this type is not intended to duplicate the processes used to make genuine banknotes, but it instead simulates the result with much-less-expensive equipment. This type of counterfeiting no longer requires artisans to engrave intaglio plates, nor does it require a large investment. Advances have put the technical means to counterfeit in the hands of ordinary people who, if they so choose, can manufacture a few counterfeits on an irregular basis. The ubiquity of home computers means that casual computer users can now more easily make high-quality simulations of banknotes and their features. It is this trend that led to the most recent major redesigns of U.S. banknotes. As part of that process, the BEP has asked the National Research Council (NRC) to undertake several studies over the past 20 years to evaluate the emerging counterfeiting threat posed by advanced reprographic technology. Box 1-2 discusses some of the most important outcomes from those reports.

COUNTERFEITING AT HOME AND ABROAD

While U.S. dollars are recognized worldwide as legal tender, their distribution and use vary in some interesting ways. Currently, approximately $720 billion in U.S. banknotes are in worldwide circulation. This amount is increasing by about 6.5 percent per year.[8] The rate of counterfeiting of U.S. banknotes is estimated to

[7]Founded in 1865 to suppress the widespread counterfeiting of U.S. currency, the U.S. Secret Service maintains exclusive jurisdiction for investigations involving the counterfeiting of U.S. currency.

[8]E. Foster, Federal Reserve Board. 2005. Presentation to this committee, May 24.

BOX 1-2
Conclusions from Previous National Research Council Reports on Currency

Three National Research Council reports on currency provide the following general conclusions regarding the changing counterfeiting threat:

- The potential threat to the United States currency from modern reprographic technology is great, due primarily to the expected increase in availability of high-quality color copier and scanner-printer combinations during the next five years.[1]
- A broadening of the counterfeiting base made possible by the availability of commercial reprographic equipment can pose an intractable enforcement problem and cause serious erosion of confidence in United States currency.[2]
- Rapid developments in reprographic technology could give rise to an unacceptable level of counterfeiting activity by making high-quality reprographic systems widely available.[3]
- The increased availability of advanced color copiers and systems composed of a computer-scanner and printer makes widespread counterfeiting of U.S. banknotes a real and substantial threat. Ready access and ease of use could lead to abuse by "casual" counterfeiters. Copiers certainly pose a significant threat, but the most important threat in the foreseeable future . . . is color scanner-computer-printer systems, aided by the continuing evolution of more-sophisticated image-processing software. These systems also provide additional opportunities for professional counterfeiters.[4]

These studies identified potential counterfeiting threats posed by technologies that primarily replicate the visual appearance of banknotes. Until recently, most casual counterfeiters have focused on reproducing the visual appearance of a banknote while using primitive methods to replicate other features, such as the banknote's tactile properties. However, emerging technology is being extended beyond the image-reproduction capability to the capacity for simulating or duplicating tactile and other nonvisual features. Also, new consumer products in crafting supplies, automotive touch-up paints, and nail polish are a few of the tools, along with new technologies, that can provide the counterfeiter with ways to replicate both the look and feel of the U.S. banknote.

[1]National Research Council. 1985. Advanced Reprographic Systems: Counterfeiting Threat Assessment and Deterrent Measures. Washington, D.C.: National Academy Press.
[2]National Research Council. 1985. Advanced Reprographic Systems: Counterfeiting Threat Assessment and Deterrent Measures. Washington, D.C.: National Academy Press.
[3]National Research Council. 1987. Counterfeit Threats and Deterrent Measures. Washington, D.C.: National Academy Press.
[4]National Research Council. 1993. Counterfeit Deterrent Features for the Next-Generation Currency Design. Washington, D.C.: National Academy Press.

be 5 counterfeits per million notes circulating in 2002. As expected, however, more \$100 notes than \$1 notes are counterfeited; the number of \$100 notes is higher than the average; it is estimated at 30 counterfeits per million.[9]

[9]J. Haslop, De La Rue. 2005. Presentation to this committee, July 21.

Approximately two-thirds of the value of U.S. currency in circulation and 70 percent of all $100 notes in circulation are estimated to reside overseas, in both dollarized and nondollarized countries.[10] In many foreign countries, U.S. currency is accepted in transactions as a global currency; it can serve as a stockpile against political and economic uncertainty; and it can provide a stable, anonymous liquid asset to individuals and corporations. The remaining one-third of the value of U.S. currency is held domestically.[11] In contrast to foreign holdings of U.S. currency, domestically held currency primarily circulates in the economy rather that being held long term.

Counterfeits are discovered in two main ways:

- Counterfeiters are found through law enforcement procedures and their products are seized, usually in large quantities.
- Merchants, banks, or private citizens discover counterfeit notes after they have been passed into circulation.

Given the global distribution pattern of U.S. banknotes, it is not surprising that the most-counterfeited U.S. note abroad is the $100 note. Because foreign cash handlers generally screen U.S. banknotes more carefully than do U.S. cash handlers, the annual dollar value of passed counterfeit notes reported abroad is constant and small, although a precise number is not reliably measured.[12] The high scrutiny paid abroad to U.S. notes means a higher passing threshold, requiring a higher investment from the counterfeiter to produce a reasonably high-quality product. Typically this threshold necessitates traditional methods such as offset or intaglio printing, high-quality paper, and reasonable simulations of security features—and large operations.

The U.S. anticounterfeiting strategy overseas, therefore, has successfully focused on enforcement. Seizures of counterfeit banknotes in foreign countries decreased from $350 million in 1995 to $20 million in 2005 as the U.S. Secret Service succeeded in shutting down large counterfeiting operations, particularly in Colombia (reportedly the source of 40 percent of counterfeit U.S. currency) and as diplomatic efforts curtailed operations in Bulgaria.[13] Other large producers of counterfeit U.S. currency are known to be in countries including Mexico, Nigeria, and North Korea.

---

[10]E. Foster, Federal Reserve Board. 2005. Presentation to this committee, May 24. Dollarized economies are those that have adopted the U.S. dollar as their official national currency.

[11]L. DiNunzio and L. Clarke. 2004. The new color of money: Safer, smarter, more secure. Proceedings of SPIE [International Society for Optical Engineering], Optical Security and Counterfeit Deterrence Techniques V, R.L. van Renesse (ed.), Vol. 5310, pp. 425-439.

[12]L. Felix, Bureau of Engraving and Printing. 2005. Presentation to this committee, May 24.

[13]L. Felix, Bureau of Engraving and Printing. 2005. Presentation to this committee, May 24.

TABLE 1-1 Amount of U.S. Banknote Counterfeiting (in U.S. dollars) in Fiscal Year 2005, by Type of Counterfeit

Type of Counterfeit	Total Amount ($)
Domestic passed	56,228,478
Domestic seized	14,680,798
Foreign passed	4,797,377
Foreign seized	20,341,615

SOURCE: Data provided to this committee by the U.S. Secret Service.

As compared with foreign counterfeiting of U.S. banknotes, domestic counterfeiting is considerably more opportunistic, is generally smaller in scale, and focuses on smaller bill denominations. In the United States, the $20 note is the most widely used note, primarily because it is the note most commonly distributed through automated teller machines, or ATMs. It is also the most counterfeited note domestically. According to data provided to the committee, counterfeit seizures in foreign countries outstrip the number of U.S. dollar notes passed in foreign countries by a factor of four (see Table 1-1). However, the reverse is true in the United States. Domestic counterfeits tend not to be stockpiled, and in the United States passed notes exceed seized notes by a factor of four. The committee also learned that a sizeable portion of counterfeits passed in the United States are produced abroad.[14]

Counterfeiting technology has been following digital reproduction technology trends. In 1995, for example, less than 1 percent of counterfeit notes detected in the United States were digitally produced. By 2005, that number had grown to nearly 35 percent worldwide and 54 percent (ink-jet printing and electrophotography) within the United States (see Table 1-1).[15] At present, ink jet is the primary technology that can easily and cheaply simulate the look of a current Federal Reserve note. The widespread availability of a variety of ink-jet printers means that notes may be printed in widely dispersed locations on a very irregular schedule and may be almost impossible to track. Chapter 3 discusses the current and future technology threats to U.S. currency in more detail.

The total dollar value of domestically passed notes is around $40 million to $50 million annually and has been approximately constant over time.[16] Because Federal Reserve machine readers capture all counterfeits that pass through Federal Reserve banks, this number is a good lower bound of the counterfeiting activity

[14]L. DiNunzio and L. Clarke. 2004. The new color of money: Safer, smarter, more secure. Proceedings of SPIE [International Society for Optical Engineering], Optical Security and Counterfeit Deterrence Techniques V, R.L. van Renesse (ed.), Vol. 5310, pp. 425-439.

[15]L. Pagano, U.S. Secret Service. 2005. Presentation to this committee, May 24.

[16]L. Felix, Bureau of Engraving and Printing. 2005. Presentation to this committee, May 24.

in the United States. However, the estimate does not include counterfeit notes that are withdrawn from circulation by recipients who neither report them nor pass them on to others.

An effective defense against counterfeiting requires an examination of technological advancements that can be employed by potential counterfeiters. It is clear that distributed, low-volume, casual counterfeiting poses a significant challenge to law enforcement. Indeed, much of the BEP's rationale for upgrading the security features in U.S. banknotes has been based in the countering of this threat. The resulting proactive analysis has significantly influenced the design changes and incorporation of additional security features in U.S. banknotes during the past 20 years.

The continual development of technologies that are exploited for counterfeiting make necessary a regular assessment of the current state and near-term outlook for these technologies and of the threats that they may pose in terms of providing capabilities for creating counterfeit FRNs. These trends also imply that more elaborate optical and substrate features will need to be incorporated in the FRNs of the future in order to stay ahead of counterfeiters. The increase in digital counterfeits has also increased the availability of point-of-use machine detection of counterfeits. This may someday be a viable means to detect both counterfeit notes and their sources in an era of greater and more widespread use of digital image technology.

The counterfeiting threat is a system composed of several processes that combine to influence the amount of counterfeit notes in circulation. Understanding and addressing the growing complexity of technologies used to produce, to verify, and to counterfeit FRNs require an integrated model that permits a synthetic view and analysis of the application of new technology not only in the creation of counterfeit notes but also in their detection, their removal from circulation, and the identification of their sources. Developing a model, even a semi-quantitative one, for this system can highlight where to allocate resources and how to design notes with a combination of features to combat counterfeiting in an optimal manner. Chapter 2, "Understanding Counterfeiting," presents the committee's findings on counterfeiting statistics for U.S. FRNs, describes five classes of counterfeiters, and summarizes a flow model of counterfeiting that can be further developed to provide quantitative insight into the deterrence effectiveness of new and current features.

THE IMPACT OF COUNTERFEITING

Counterfeiting can have a number of kinds of impact on individuals, companies, financial institutions, and the nation that issues the currency. The most obvious impact might be expected to be economic. However, the total value of counterfeit notes passed (about $61 million in 2005) is less than 0.01 percent of the value of currency in circulation and an even smaller portion of the total U.S. economy.[17] Counterfeiting of banknotes is very small compared, for instance, with counterfeiting of credit cards or of branded goods. The counterfeited-products business engenders huge losses, ranging from 5 to 8 percent of worldwide sales of brand products, and credit card fraud accounted for more than $750 million in losses in the United States in 2004.[18]

Good design and strong enforcement policies have enabled U.S. currency to achieve one of the lowest rates of counterfeiting of any major currency. For all notes the rate is only 5 notes per million, but this rate is low owing in part to the large number of $1 notes in circulation. The number of $100 notes counterfeited runs at about 30 counterfeits per million. By comparison, the euro is estimated to be counterfeited at present at 65 notes per million, the British pound at 160 notes per million, and the Canadian dollar at 1,000 notes per million.[19]

Counterfeiting can have psychological effects and, as a result, governments (and their enemies) have long realized that counterfeiting is a national security issue. An example can be found in a British government attempt to destabilize the Continental government by counterfeiting U.S. currency during the Revolutionary War.[20] Similarly, the Union government sent counterfeit Confederate dollars south during the Civil War,[21] and during World War II, the German government manufactured a high-quality counterfeit of the British £5 note.[22] A further illustration of the psychological impact of counterfeiting is provided by the Canadian experience of recent counterfeiting. The Bank of Canada reports that when the level of counterfeit Canadian $100 notes reached 300 notes per million, as many as 11 percent of merchants stopped accepting $100 notes.[23] Cognizant of these threats, the United States and other countries have anticounterfeiting strategies intended to maintain consumer confidence in U.S. currency, just as some politically motivated counterfeiting is intended to undermine it.

More subjective, but also a critical aspect for the United States, is the political

[17]U.S. Department of the Treasury. 2003. The Use and Counterfeiting of United States Currency Abroad, Part 2. The second report to the Congress by the Secretary of the Treasury, in consultation with the Advanced Counterfeit Deterrence Steering Committee, pursuant to Section 807 of Public Law 104-132. Available at <http://www.federalreserve.gov/boarddocs/rptcongress/counterfeit2003.pdf>. Accessed April 2006. Note that the percentages of counterfeits will vary if reported as the percentage of banknotes or the percentage of monetary value.

[18]Credit card fraud in the U.S. 2005. The Nilson Report, Vol. 830, p. 8.

[19]J. Haslop, De La Rue. 2005. Presentation to this committee, July 21.

[20]See, for example, <http://www.secretservice.gov>. Accessed February 2007.

[21]G. Tremmel. 2003. Counterfeit Currency of the Confederate States of America. Jefferson, N.C.: McFarland.

[22]Bank of England Fact Sheet. 2003. Available at <http://www.bankofengland.co.uk/banknotes/factnote.pdf>. Accessed February 2007.

[23]J.F. Chant. 2004. Bank of Canada Review. Ottawa, Ontario: Bank of Canada. P. 43.

advantage to issuing a global currency, along with the threat of that advantage being undermined if there is a perceived threat to the security of the U.S dollar. The loss of U.S. prestige and influence abroad owing to any lack of confidence in the dollar is not readily measurable, but it would nonetheless be severe.

Although the total economic cost of counterfeiting is low, the personal cost to those left holding a worthless note may be high. When a counterfeit note is identified, it loses its value. When this cost accrues to an innocent party—usually a retailer or service provider—it is in the best interest of any government to protect its citizens from this threat.

Laws of the United States prohibit the spending or possession of counterfeits and require the reporting of the receipt of a counterfeit banknote. Reported counterfeits generally produce a direct monetary loss to those who receive them because the fake bills must be surrendered to authorities without compensation.[24] With no compensation incentive, a pattern of complacency may have emerged, with many people and businesses willingly absorbing the monetary loss of the value of the counterfeit rather than taking the trouble to report it. Many would choose to avoid what could be an intrusive and time-consuming interaction with authorities. The potential awkwardness of reporting a genuine note as a fake is also present.

DETECTION OF COUNTERFEITS: THE EFFECTIVENESS OF FEATURES ON U.S. BANKNOTES

The effectiveness of a banknote feature depends on the specific nature of a note's use. Examples include different types of use by individuals wishing to complete a sale—including use by the blind—and transactions by means of machines designed to accept, dispense, or count banknotes. Banknote design is complicated by currency's wide variety of uses. A given feature on a banknote may address just one or many of them. The wide range in types of use for FRNs is complicated further by the fact that these transactions may take place anywhere on the globe under a wide range of ambient lighting conditions.

For any feature to be effective, it must work for the duration of the life of the note. A variety of durability tests are conducted by the BEP to ensure that banknotes have a reasonable circulation life. Tests include repeated crumpling and folding, soiling and occasional laundering of notes, as well as tests of their wet tensile strength, lightfastness, and chemical resistance to a variety of fluids. Interestingly, the durability of banknotes is itself a security feature. Experts report that counterfeit notes are often identified by their poor and uneven wear.[25]

---

[24]Some home insurance policies insure against losses due to accepting counterfeit currency.

[25]J. Haslop, De La Rue. 2005. Presentation to this committee, July 21.

Human Detection of Counterfeits: Visual and Tactile Effectiveness

The most striking visual feature of an FRN is the portrait on the front of the note. It is the largest single feature, and because it is a human face, many people find it easy to recognize small changes in its proportions or coloring.

Each note also has one large, high-contrast numeral for use in low-light environments and by the visually impaired. However, this feature and others do not provide adequate differentiation for many visually impaired individuals and provide no method of differentiation for blind persons.[26,27]

Overt visual features on U.S. banknotes are designed to require only direct visual inspection for authentication. Many are also intended to be impossible to replicate using low-resolution computer technology. These include the watermark images, embedded plastic security strips, and color-shifting ink. Each of these features is denomination-specific; the watermarks repeat the portrait on each denomination, and the security strips are embedded in a different position and contain different text and graphics for each denomination.

Perhaps more important than the effectiveness of individual features is the effectiveness of the combination of features on the entire note. The placement of features and how they interact, however, is difficult to gauge. For example, printing over the many visual security features in the paper substrate may make them less effective. Crowding the note with too many features may result in users not noticing any of them.

The lack of comprehensive, scientifically rigorous quantitative information on the effectiveness of individual features makes feature selection difficult. Specifically, a variety of systematic and well-designed statistical tests would have been very useful in decision making regarding recent feature modification and implementation.

The BEP conducted a focus-group study in 2001 and 2002 to determine how specific cash-handling audiences detect counterfeits in Series 1996 banknotes.[28] In the course of this study, the bureau conducted interviews of a total of 1,423 people in six categories: consumers, bank tellers, cashiers, gaming industry employees, law enforcement officials, and teachers. It was found that 29 percent of the cash

[26]National Research Council. 1995. Currency Features for Visually Impaired People. Washington, D.C.: National Academy Press.

[27]Information at <http://www.ourmoneytoo.org/position.php> refers to strategies of individuals and coalitions for improving this aspect of U.S. banknotes. Accessed February 2007.

[28]L. DiNunzio and S.E. Church. 2002. Evaluating public awareness of new currency design features. Proceedings of SPIE [International Society for Optical Engineering], Optical Security and Counterfeit Deterrence Techniques V, R.L. van Renesse (ed.), Vol. 4677, pp. 1-14. Also, H. Treinen. 2004. Banknote recognition: Public reception of double sided intaglio printed banknotes compared with single sided intaglio and offset counterfeited banknotes. Currency Conference, Rome, 2004. Personal communication.

handlers interviewed had identified counterfeit notes; the major tip-off was that the note "looked suspicious," followed closely by the fact that it "felt suspicious." In order, the features that caused interviewees to look closer were these: color, waxy feel, paper feel, paper thickness, smudged ink, portrait quality, and security strip. Color is the first feature checked, according to the group members interviewed, and color is now easily duplicated by commonly available computer technology. Thus, the new series of notes may become less secure because cash handlers might stop looking for other features if the color appears to be "correct."

The BEP focus-group members reported that when they suspected a counterfeit, the features that they used for confirmation, in order, were these: watermark, starch content (indicated by use of a special pen), security strip, feel, color-shifting ink, fine lines, and shading in the denomination number. The features that this focus group was most aware of were, in order: watermark, security strip, color-shifting ink, fine lines, and microprinting.

Although interesting and informative, focus-group data are not as useful as those obtained in controlled studies, and self-reported usage of features should be interpreted with caution. In more controlled studies on U.S. currency, researchers have observed that "experiments indicate that people are good at detecting counterfeits, that inkjet counterfeits are easier to detect than offset counterfeits, and that counterfeits of the newly designed bills are easier to detect than counterfeits of the older series. The design improvement was greatest with the $100 bills and, to a lesser extent, $50 bills."[29]

A marked improvement was also noted in the recognition of the copper-to-green color-shifting ink now in use on the $10, $20, and $50 notes versus the original green-to-black ink on the $10 and $100 notes. However, a key observation made in the course of this study was that "judging the improvement of features was not the same as judging their absolute efficacy."[30] When asking participants in the study to look at single features on the notes, the researchers noticed a variation in perception between situations in which an entire note was presented and those in which the rest of the note was masked and only a single feature was visible. Experience and focus were also noted as key discriminators.

Studies of banknote features on other nations' currencies have confirmed the importance of subjective considerations, including the consistency of the feature

---

[29]A.P. Hillstrom and I.H. Bernstein. 2002. Counterfeit detection for new and old currency designs. Proceedings of SPIE, Optical Security and Counterfeit Deterrence Techniques IV, R.L. van Renesse (ed.), Vol. 4677, pp. 65-80.

[30]A.P. Hillstrom and I.H. Bernstein. 2002. Counterfeit detection for new and old currency designs. Proceedings of SPIE, Optical Security and Counterfeit Deterrence Techniques IV, R.L. van Renesse (ed.), Vol. 4677, pp. 65-80.

TABLE 1-2 Committee's Analysis of Banknote Features Used in Commercial Machine Authentication

	Class of Machine Reader				
	Single-Note Denominator				
Feature	Low End	High End	Single-Note Authenticator	Desktop Counter	High-Speed Counter-Sorter
Optical spectrum		●	●	●	●
Magnetic properties		●	●	●	●
Paper fluorescence			●	●	●
Fluorescent strip			●	●	●
Color-shifting inks	●	●	●		

NOTE: Features not used in commercial machine authentication include the watermark, micro-printing, portrait, freedom symbols, intaglio printing, color, digital counterfeit-deterrence system, metallic ink seals, and colored fibers. "●" indicates use in commercial machine authentication; "blank," not used.

from note to note and the complexity of the overall design.[31,32] All of the studies cited observe that education of the public, with emphasis on cash handlers, plays a large role in the recognition of a genuine note.

Machine Counting and Authentication

In addition to the visually and tactilely detectable features of FRNs—a key line of defense against the passing of counterfeit notes—several machine-readable features are useful for both authentication and denomination. Machine reading of banknotes is a new technology; it was virtually nonexistent 15 years ago but now annually processes more than $64 billion in U.S. currency. The features of U.S. banknotes most used in current machine readers are the optical spectrum and image, magnetic inks, ultraviolet fluorescence, ultraviolet spectrum, and infrared ink pattern. Low-end readers may sense only a single feature, usually the infrared ink pattern; high-end readers may use 10 or more measurements to authenticate each note.

Four classes of machine reader are shown in Table 1-2. Each of these machines relies on different sets of banknote features:

[31]A.A. Andrade. 2004. Assessing the security of a hologram with the assistance of a multi-criteria decision analysis. Keesing Journal of Documents and Identity 9:10-14.

[32]R.M. Klein, S. Gadbois, and J.J. Christie. 2004. Perception and detection of counterfeit currency in Canada: Note quality, training and security features. Proceedings of SPIE, Optical Security and Counterfeit Deterrence Techniques V, R.L. van Renesse (ed.), Vol. 5310, pp. 1-12.

- *Single-note denominators.* Commonly found in vending machines and change machines and at self-checkout stations, these typically use infrared, broad-wavelength optical, or magnetic sensors to detect denomination-specific features.
- *Single-note authenticators.* These typically include additional sensors for the detection of ultraviolet and fluorescent patterns and for the identification of individual features.
- *Desktop counters.* These are used to sort and count large numbers of notes at high speeds, up to thousands of notes per minute. They typically employ broad-wavelength optical imaging, ultraviolet spectrum, and magnetic signals.
- *High-speed counter-sorters.* These require features that give a strong signal that is not highly position dependent; thus, neither the ultraviolet and infrared ink features nor the gamut of overt features are generally sensed. Large-scale counter-sorters, used typically by banks, casinos, and high-volume businesses, employ detection technologies similar to those of desktop counters.

Typically, manufacturers of machine readers report that low-quality counterfeits are identified by a low optical image quality, lack of magnetic and/or infrared ink, or incorrect paper fluorescence. It may require detailed magnetic signature sensing or ultraviolet spectrum sensing to detect high-quality counterfeits. Most overt counterfeit-deterrent features are not used by machine readers because, as explained below, they would be difficult for these machines to sense, locate, or verify.

Typical machine readers use point sensors that scan a narrow band as the note moves through the reader; they do not scan the full width of the bill. To do so would require an array of, or rastered, point sensors, a design that would significantly increase cost owing to the instrumentation and data-processing requirements. The design compromise thus used in current machine readers makes small, distinct features such as security strips, which occur in different locations on each denomination, difficult to use because the sensor "window" might miss them. In addition, the large data set collected by scanning the complete note would require too much time to analyze. Because point sensors are used, image-recognition schemes, which require scanning a large area, are not practical. Finally, features that move when a note is redesigned may move out of the sensor window.

For a high-speed counter, the signal strength from each note must be high enough to be sensed in 0.04 second. Several features can be useful for reading in this fraction of a second, including the magnetic ink signal, the pattern in infrared ink, and the ultraviolet spectrum and fluorescence of the paper. High color contrast, as in previous, all-intaglio note series, provides a strong optical signal; however, artistically smooth shadings and the addition of multiple colors and features cause the optical contrast to decrease. In addition, for the purposes of machine readers, the signal from a feature must be highly reliable. Overall color, which changes with use, is an example of an unreliable signal.

Machine readers feed notes in one of two directions. "Short-end-first" readers include nearly all single-note readers, as well as some large-scale counter-sorters. All other high-speed readers take notes "wide-end first." The advantage of short-end-first readers is that sensing the length of the bill provides more information and a higher signal. Wide-end-first readers have the advantage in speed because they read a shorter path per bill.

Certain features on current notes are easily read in either feed direction. These include the ultraviolet spectrum and fluorescence of the paper and the patterns in magnetic and optical ink in the printed image. An example of a feature that is not readable in both directions is the infrared ink pattern, which is a set of stripes parallel to the short edge of the note.[33] A short-end-first reader senses an on-off pattern as the counter moves along the note, whereas a wide-end-first reader cannot sense the stripe pattern, but only whether the sensor is or is not within a stripe.

CONCLUSIONS

- Although U.S. currency has one of the lowest rates of counterfeiting of any major currency, counterfeiting remains an economic, psychological, and political threat. Therefore, it is in the national interest to preserve the actual and perceived security of U.S. currency.
- Today, domestic counterfeiting—dominated by digital imaging technologies—focuses on the $20 note. Foreign counterfeiting centers on the $100 note produced by more traditional methods. It is likely that the increasing accessibility of advanced printing equipment, materials, and information will blur the distinctions between classes of counterfeiters.
- Security features that maintain the "look and feel" of historical U.S. banknotes have been added to today's Federal Reserve notes. These new features—including security strips, watermarks, embedded fibers, and color-shifting ink, as well as microprinting, fine-line printing, and color printing—provide means for counterfeit deterrence and authentication as well as presenting difficulties for nonauthorized sources attempting to replicate the notes.

---

[33]Staff members, Cummins-Allison Corporation, Mount Prospect, Ill. 2005. Discussions during a subcommittee visit, October 7.

- The security features in current use are highly durable, low-cost, odorless,[34] and environmentally sound. Many of the features are detectable by the unaided eye. The unique look and feel of the substrate itself is an important part of the FRN's recognizability, so printing over much of it may be counterproductive.
- The use of machine readers for currency is increasing worldwide; the security features used by machine readers differ from those used by human cash handlers. Because the machine-readable features of currency are sometimes changed as currency design changes, the design of these machines must be modified with each currency design change. Additionally, the orientation of features may not necessarily work optimally with high-speed machine feeders; such incompatibility can limit the machine's functionality.

[34]It is interesting to note that banknotes are odorless once the ink has fully dried; however, some volatiles may be detected by scent in brand-new notes.

2

Understanding Counterfeiting

Examining the counterfeiting threats from technology and developing novel counterfeit-deterrence features for future implementation require a full understanding of recent trends in counterfeiting and of the evolving characteristics and competences of counterfeiters. Since the committee, like the majority of this report's expected readership, was not composed of currency experts, it concluded that two important steps in determining how to deter identified counterfeiting threats are to understand (1) who is doing the counterfeiting and (2) what the different steps in the counterfeiting "system" are. These issues proved to be useful tools for evaluating the effectiveness of potential features.

PORTRAIT OF A COUNTERFEITER

The portrait of a counterfeiter is intrinsically linked to the evolution of the technology of currency production and the evolution of technologies that can be used to replicate and simulate currency features. Five classes of counterfeiter can be characterized by the technologies and the methods that they use—primitive, opportunist, petty criminal, professional criminal, and state-sponsored. Table 2-1 summarizes the key attributes of each class, and Table 2-2 characterizes their activities and methods of passing notes. Table 2-3 shows what access to various kinds of current digital technology the different classes of counterfeiter have. The relative threats associated with each criminal class are depicted in Figure 2-1. These portraits are expanded on below.

TABLE 2-1 Classes of Banknote Counterfeiter, Their Tools, General Location, and Impact

Class	Typical Practitioner	Primary Tools	Location	Impact of Activity
Primitive	Unusually motivated individual	Manual artistry	Domestic or foreign	Very low
Opportunist	Opportunistic young adult, typically works alone	Home office equipment	Domestic	Created largest increase in $20 domestic-passed currency
Petty criminal	All ages, criminal intent, typically works alone	Home office equipment plus specialty materials and processes	Domestic	Low, stable level of activity
Professional criminal	Criminal, trained in printing technology, often part of a criminal group	Offset printing, high-end ink-jet printers, specialty materials and processes	Domestic or foreign	Low, stable level of activity
State-sponsored	Professional, profiteer or terrorist, member of a large organization	All materials and processes, including specialty paper, intaglio and offset printing, security features	Foreign	Strategic concern

TABLE 2-2 Methods and Extent of Dissemination of Counterfeit Banknotes, by Class of Counterfeiter

Class	Production Level	Stockpiling	How Notes Are Passed
Primitive	Very small	None	Individually, by counterfeiter
Opportunist	Small, as needed	None	Individually, by counterfeiter or friends
Petty criminal	Small to moderate, often over years	None to moderate	Individually, by counterfeiter or criminal associates
Professional criminal	Large	Large	Through criminal networks
State-sponsored	Large	Unknown, presumably large	Through various legitimate and illegitimate networks, often by unwitting accomplices

TABLE 2-3 Digital Technology Access, by Class of Counterfeiter

Class	Ink-Jet Printer	All-in-One Device	Color Copier	Flatbed Ink-Jet Printer	Digital Press	High-Quality Scanner	Imaging Software
Primitive	Not applicable—does not use digital technology.						
Opportunist	●	●				●	●
Petty criminal	●	●	●			●	●
Professional criminal	●	●	●	●	●	●	●
State-sponsored	Not applicable—reproduces government processes directly.						

NOTE: Within all counterfeiter classes, additional nondigital techniques may be used to improve note simulations (for example, craft supplies to reproduce features that use color-shifting ink). "●" indicates high likelihood of access; "blank," low likelihood of access.

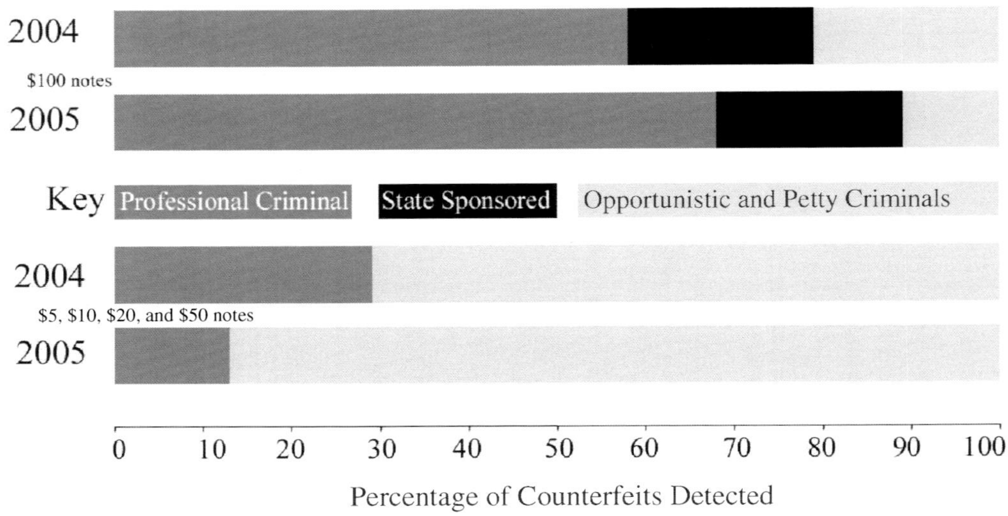

FIGURE 2-1 Comparison of the percentages of counterfeit notes—created by professional criminals, state-sponsored counterfeiters, and opportunist and petty criminals—detected in 2004 and 2005. The two upper bars represent the percentage of $100 counterfeits by source and show the primary source to be professional criminals. The two lower bars represent percentages by source for the sum of $5, $10, $20, and $50 counterfeits, showing the growing dominance of opportunist and petty criminals (the light-grey segments). Note that state-sponsored counterfeits are identified only for $100 notes (the black segments) and that primitive counterfeiters do not account for enough counterfeits to appear on this chart. SOURCE: Data provided to the committee by the U.S. Secret Service.

The Primitive Counterfeiter

The criminal perpetrator of primitive counterfeiting may use manual artistry to modify a piece of currency in order to increase its value and obtain financial gain. The phony notes thus produced are easily detected by attentive cash handlers and, upon examination, by the general public. An example is a note that has cut-and-pasted numbers increasing the denomination of a $10 bill to $100.

The primitive counterfeiter's products are often obvious to the point of parody. They are clearly incompatible with automatic currency-authentication equipment and must be passed person to person. If the substrate is an existing Federal Reserve note (FRN), the counterfeit note retains the correct feel of genuine currency and may have advanced security features such as a watermark and a security strip, albeit for the wrong denomination.

Such a note may be passed in a number of ways—for example, the counterfeiter may try to distract the cashier to prevent scrutiny of the counterfeit. He or she may also rely on the reluctance of a cashier to question the authenticity of a banknote in a point-of-sale situation.

The Opportunist Counterfeiter

Over the past decade, desktop publishing has become the dominant tool for counterfeit printing operations in the United States. In fiscal year 2001, 93 percent of suppressed counterfeiting operations used digital processes; this is a very significant increase, from less than 20 percent in 1995.[1] This trend, begun in the United States, has continued worldwide. Digital publishing tools, including color copiers, scanners, and ink-jet printers—all typically found in a college dormitory room or home office—powered by information available on the Internet, are the engines that impel opportunist counterfeiters to undertake this criminal activity. Increasingly, opportunist counterfeiters are also using all-in-one, combination scanner-and-printer devices. The devices may be controlled by a computer with image-processing software to modify the image and connect to the Internet. Instead of learning from master engravers and operators of offset printers, the opportunist uses Internet searches to obtain know-how and may tell no one else or might share the information with a few friends. The designation of "opportunist" implies that this counterfeiter uses only equipment that is commercially available and that has been obtained for more legitimate uses.

Opportunist counterfeiters are typically young adults, with a median age of perhaps 18 years. They work as individuals, making only a few notes at a time and printing them "on demand" from their residence or place of employment. They typically pass these notes personally or via friends in retail transactions. Their counterfeits span a wide range in quality, from single-sided notes[2] to impressive duplications requiring many hours of refinement in touching up colors, aligning front and back images, and refining enhancements to the printed image to simulate advanced security features using arts-and-crafts supplies. The $20 note is the denomination most commonly produced by this class of counterfeiter, because it is the note expected in many transactions.

As is true for primitive counterfeits, the vast majority of counterfeits made by opportunists are detected at or near the point-of-sale. They rarely are passed more than once and generally are discovered before circulating back to a Federal Reserve Bank (FRB). Eighty percent of these counterfeits are turned over to the U.S. Secret Service by commercial establishments, financial institutions, and law enforcement.[3]

---

[1]U.S. Department of the Treasury. 2003. The Use and Counterfeiting of United States Currency Abroad, Part 2. The second report to the Congress by the Secretary of the Treasury, in consultation with the Advanced Counterfeit Deterrence Steering Committee, pursuant to Section 807 of Public Law 104-132. P. 60. Available at <http://www.federalreserve.gov/boarddocs/rptcongress/counterfeit2003.pdf>. Accessed February 2007.

[2]How Counterfeiting Works. Available at <http://money.howstuffworks.com/counterfeit.htm>. Accessed February 2007.

[3]United States Department of the Treasury. 2000. The Use and Counterfeiting of United States Currency Abroad. Report to the Congress by the Secretary of the Treasury, in consultation with

The opportunist's typically youthful age and exploratory approach often lead to situations with unexpected consequences. For these reasons, widely informing the public of the U.S. Secret Service's conviction rate—higher than 90 percent[4]—of counterfeiters may significantly reduce the opportunist's temptation to "make a little money."

The Petty Criminal Counterfeiter

The petty criminal counterfeiter has a clear and systematic intent to counterfeit repeatedly. These practitioners use the same digital tools that opportunists employ and may supplement their efforts with specific materials such as the best paper and inks. Using the Internet, they develop and share a set of tricks to improve their simulations and to help them pass their phony bills. The tricks include methods to bypass or negate certain authentication methods, such as the "iodine" starch-detecting pen, as well as ways to handle the encounter when their creation is questioned. This type of operation is still an individual or small effort, but it differs from the activities of the opportunist in duration, quantity, or distribution area. Some of the best counterfeits in this class display innovative ways to simulate the security strip, the watermark, and the specialty inks. This class of counterfeiter will likely be the first to challenge automatic currency authenticators.

The Professional Criminal Counterfeiter

The professional criminal counterfeiter class generates phony bills that are typically relatively easy to pass to the public. These counterfeiters simulate all of the critical features of genuine notes to some degree—in fact, the simulations often require additional criminal activity in order to acquire controlled materials such as security inks and paper. These counterfeiters are typically part of a larger criminal enterprise that can include dedicated specialists who sometimes have professional training in the printing business. Their efforts involve generating significant quantities of counterfeits, along with developing the necessary distribution methods—which may also be tied to other criminal endeavors, such as the counterfeiting of identification cards or other security documents. This type of distribution far surpasses that of the petty criminal.

The Secret Service actively pursues these criminal organizations, tracking their activity by classifying recovered counterfeits into groups. Each note received by the

---

the Advanced Counterfeit Deterrence Steering Committee, pursuant to Section 807 of Public Law 104-132. Available at <http://www.ustreas.gov/press/releases/reports/counterfhp154.pdf>. Accessed February 2007.

[4]L. Pagano, U.S. Secret Service. 2005. Presentation to this committee, May 24.

Secret Service is characterized, classified with the telltale information that reveals the source, and put into a database so that law enforcement and commercial banking organizations around the world can assist in counterfeit identification.

Much professional counterfeiting activity is located outside the United States, although a substantial fraction of the counterfeit notes may be passed domestically. The target denominations of professionals criminals include the $20, $50, and $100 notes—with the $100 note being the most counterfeited overseas. Judging by the typical amount of currency seized when the government shuts down professional operations, this class of counterfeiter can pose a serious threat. However, continuous enforcement efforts by the U.S. government, often working with foreign governments to change and enforce laws, as well as attention to foreign currency-handling procedures and the introduction of the new currency designs, have kept the impact from this potentially significant source of counterfeits low.

The State-Sponsored Counterfeiter

State-sponsored counterfeiters not only plan for criminal financial gain but may also have a political goal—such as reducing global confidence in U.S. currency. Thus, they are willing to invest in technologies to duplicate U.S. banknote features exactly. State-sponsored counterfeits may be passed by both apparently legitimate and criminal means. The very-high-quality so-called supernotes that duplicate nearly all of the security features of genuine banknotes are created by this class of counterfeiter.

Some state-sponsored organizations make their own paper with watermarks and colored threads, make their own specialty inks, and re-author the engraved image. They use the same printing methods—intaglio and letterpress—and may integrate their own forensic features that would enable their own internal discrimination. While these notes would fool most cash handlers and even some machine authenticators, they can still be identified by the Federal Reserve System and the Secret Service. The U.S. government has confirmed its suspicion that the North Korean government is involved in the state-sponsored production of supernote counterfeits.[5] It has been reported in the media that more than $45 million is estimated to have been passed by this source since 1989.[6]

---

[5]See <http://www.state.gov/p/inl/rls/nrcrpt/2006/vol2/html/62144.htm>, <http://losangeles.fbi.gov/dojpressrel/pressrel06/la041906usa1.htm>, and <http://www.usdoj.gov/criminal/press_room/speeches/2005_4193_rmrksOprSmokngDrgnNroylChrm082405O.pdf>. Accessed February 2007.

[6]B. Gertz. 2005. N. Korea charged in counterfeiting of U.S. currency. Washington Times. December 2. Available at <http://www.washtimes.com/world/20051201-103509-5867r.htm>. Accessed February 2007.

Who Does What?

Forensic science can identify common characteristics among discovered counterfeit notes that can be traced to unique aspects of the counterfeiting process used and thereby allow the counterfeits to be classified according to their sources. Table 2-4 presents a snapshot of the magnitude of U.S. banknote counterfeiting for fiscal year 2005. These numbers are typical of the counterfeiting threat in recent years. The counterfeits are classified by the Secret Service according to their source, either domestic or foreign, and according to whether they were discovered before or after they were passed into circulation. These data provide some insights into the patterns and practices of counterfeiters.

A FLOW MODEL FOR COUNTERFEITING

Counterfeiting begins—but does not end—with the printing of bogus banknotes. After producing (and possibly stockpiling) notes of a sufficient quality and quantity, the counterfeiter still has work to do. To realize a profit from these efforts, he or she must then exchange the counterfeit notes for cash, goods, or services. After a counterfeit note has been passed, it will circulate until it is detected and removed from the system. Then, and only then, is the economic loss of counterfeiting realized: the last one holding the fake banknote loses.

Decisions about which particular feature (or combination of features) to incorporate into future FRNs must be based on many criteria: the difficulty of replication or duplication by various classes of counterfeiters, the clear ability of ordinary people to distinguish between legitimate and counterfeit notes, durability, cost, aesthetics, and so on. Gaining a consensus about how these factors should be combined into an overall figure of merit is not straightforward. The committee suggests one method to determine how some of these criteria can affect

TABLE 2-4 U.S. Banknote Counterfeiting in Fiscal Year 2005: Production Technology, Class of Counterfeiter, and Amount (in U.S. dollars)

Primary Production Technology	Primary Criminal Class	Domestic Passed ($)	Domestic Seized ($)	Foreign Passed ($)	Foreign Seized ($)
Ink-jet printing	Opportunist, petty criminal	29,153,845	5,940,531	29,984	642,841
Electrophotography	Opportunist, petty criminal	2,164,475	758,045	8,690	9,920
Offset press (domestic)	Professional	2,264,582	1,608,682	72,593	7,186,140
Offset press (foreign)	Professional	21,244,276	1,290,340	746,910	11,971,514
Intaglio press (foreign)	State-sponsored	1,401,300	5,083,200	3,939,200	531,200
TOTAL		56,228,478	14,680,798	4,797,377	20,341,615

SOURCE: Data provided to this committee by the U.S. Secret Service.

one important measure: the amount of counterfeit currency in circulation. This method uses a simplified model to map the flow of counterfeit currency from production to removal.

It is important to note that the committee does not suggest that this particular elementary model should be used to make policy decisions about adopting future features. Instead, the model is presented only as a device to provide a possible rationale for determining which features might be—in some sense—evaluated as being "better" than others. Moreover, for this (or any alternative) model to be used as an instrument for determining policy, it would have to be appropriately validated. As discussed below, this would require either estimating its parameters from existing field or experimental data, or running experiments specifically designed with this goal in mind. In any event, a truly useful model would have to be capable of considering sets of features together, since features do interact, in addition to overcoming the limitations of this quite-basic model. The model is presented as an aid in explaining the origins of the evaluation process adopted by the committee.

This said, the counterfeiting threat can be represented as a basic flow system, as shown in Figure 2-2, in which the boxes represent repositories of counterfeit currency (containing amounts measured in units of dollars). Counterfeit notes progress through the system from production, through stockpiling and, by means of a passing event, into circulation. These flows are indicated by the arrows in Figure 2-2. At each stage, disruptions to the system can be introduced by removing counterfeit notes or by deterring their production—shown as arrows pointing to the ovals that represent repositories of counterfeit currency removed from the system. The variables used to represent this system are defined in Table 2-5.

The topmost ovals in Figure 2-2 are special cases—representing counterfeits that are never actually made because of technology stoppers, or "blocks," or owing to the difficulty of feature simulation. As such, these ovals are "virtual" repositories of counterfeits; no actual notes reside there. The effect of these virtual repositories is captured in ρ, the counterfeit production rate, which reflects the effects of technology stoppers or the difficulty of feature simulation for the purpose of feature evaluation. In particular, any increased flow to the virtual repositories will necessarily decrease the production rate, ρ.

In order to best analyze this system, the parameters for each feature and for each class of counterfeiter should be quantified. While the values of some of these are known, such as u (the amount seized by the Secret Service), many others remain unknown. However, a strong case can be made for estimating or measuring these variables, since they provide a basis for quantitative analysis of the effect of new features on counterfeit deterrence, as discussed below. Even for variables that are hard to measure in the real world—that is, s, the fraction of successful pass attempts—controlled experiments could provide some insight that would aid in modeling the system. With the aid of a fully characterized flow model, such as

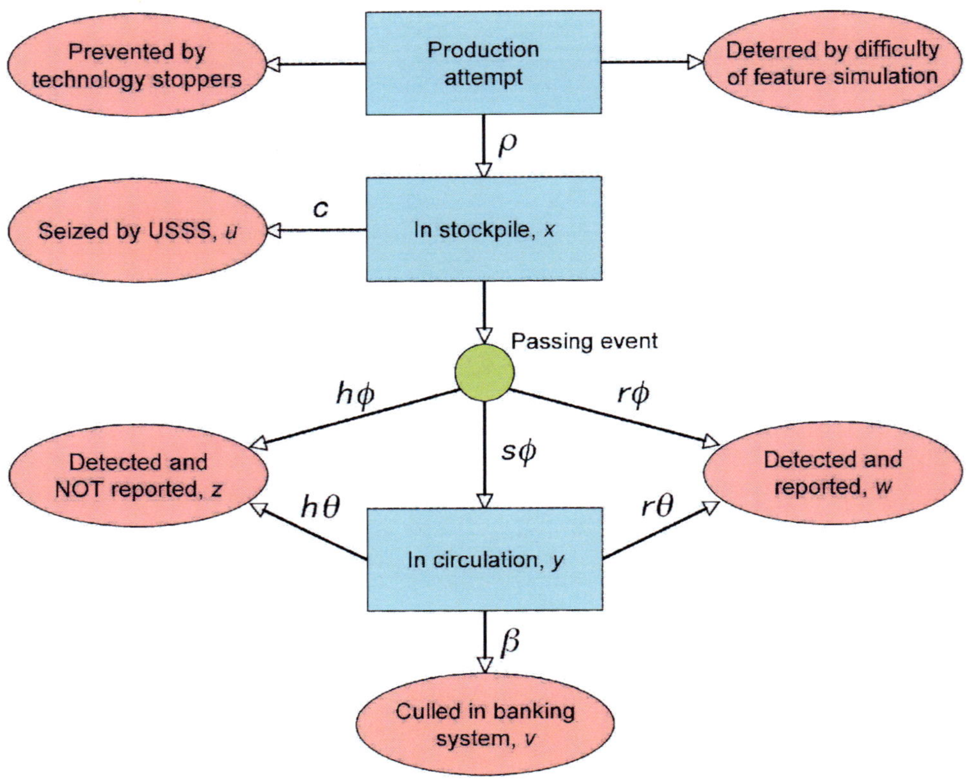

FIGURE 2-2 A flow model for counterfeiting. The boxes indicate the components of the counterfeiting system. The ovals represent processes by which counterfeit notes are removed from the system. Arrows indicate the flow of counterfeit notes. NOTE: USSS = U.S. Secret Service. Variables are defined in Table 2-5.

the simplified one presented here, or a more sophisticated version,[7] the impact of adding and subtracting features could be scientifically evaluated, and realistic cost-benefit analyses and feature comparisons could be made.

Appendix E shows the details of an analysis based on this extremely simple model and a number of simple, yet qualitatively reasonable, assumptions. The

[7]Many more realistic constructs are, of course, possible: explicitly including a nonstationary behavior of most of the parameters; having the rate of successful pass on a second attempt be different from that on the first; allowing the rate of confiscation, *c*, to be a function of the amount in the stockpile; having the rate of passing be independent of the amount in the stockpile; and so on. However, the additional burden of estimating these parameters might not improve the qualitative insight into how a particular feature might impact the amount of counterfeits in circulation.

TABLE 2-5 Variables in the Flow Model for Counterfeiting

Variables	Definition
Counterfeit repository (total $)	
u	$ seized from stockpile by U.S. Secret Service (USSS)
v	$ culled from banking system by the Federal Reserve System (FRS)
w	$ turned in to USSS by a recipient
x	$ in stockpile, awaiting first pass attempt
y	$ in circulation
z	$ held by a recipient, but not reported or repassed
Rates	
c	Confiscation rate of notes from stockpile ($ per unit time)
β	Rate of culling in the banking system by the FRS (per unit time)
ϕ	Rate at which first attempts are made to pass note from stockpile (per unit time)
θ	Rate at which repass attempts are made during circulation (per unit time)
ρ	Counterfeit production rate ($ per unit time)
Other parameters	
h	Fraction of pass attempts held and not reported to the USSS; $h = 1 - s - r$
r	Fraction of pass attempts reported by recipient
s	Fraction of successful pass attempts
t	Time since start of production

result is a set of linear differential equations, which yield the temporal change of the amounts of counterfeit currency located in the boxes and ovals, and which provide the underlying dynamical relations among the parameters and variables in the system.

The most salient output of the analysis in Appendix E with respect to feature selection is the steady-state amount of counterfeit currency in circulation given by the following equation:[8]

$$y(t=\infty) = \frac{s(\rho - c)}{(1-s)\theta + \beta} \qquad (2\text{-}1)$$

Some of the parameters in this equation may not be linked to the particulars of note features. For example, c, the confiscation rate of stockpiled counterfeits, is likely influenced most by law enforcement activities, and β, the rate of culling from the banking system, is a function of how often *all* notes pass through machine authenticators in the banking system. However, the ability of a feature to decrease the amount of circulating counterfeits is related to its ability to do the following: (1) decrease the counterfeit production rate ρ via technology stoppers or difficulty of simulation, (2) decrease the fraction of successful pass attempts s because of unacceptable feature replication or duplication, and (3) increase the rate at which repass attempts are made during circulation θ.

Perhaps the most important insight provided by Equation (2-1) is the nonlin-

[8]The derivation of this equation is explained in Appendix E of this report.

ear, nonadditive contributions of the production and pass rates to the amount of counterfeit currency in circulation. This nonlinearity, which can result in surprising conclusions, illustrates the strength of the flow model as a tool for feature analysis, as shown in hypothetical examples in Box 2-1 and in Appendix E.

The model's results, as exemplified in the Box 2-1, are salient to the committee's task of evaluating the effect of potential security features in reducing counterfeiting. Although flow models such as this have been developed for representing the

BOX 2-1
Comparison of Features Using the Flow Model

An analytical model can provide (perhaps non-intuitive) insights into the systems within which counterfeit notes circulate; such insights can then contribute to the decision-making process for the selection of promising features.

For example, consider two potential counterfeit-deterrent features: Feature A is easy to simulate visually but also easy to detect. Feature B is difficult to simulate convincingly but also difficult to detect in a casual transaction. For illustrative purposes, assume that Feature A deters only 10 percent of counterfeit attempts, but deters 90 percent of passing attempts. Feature B deters 90 percent of counterfeiting attempts, but only 10 percent of passing attempts. These numbers are reflected in the counterfeit production rate ρ and passing fraction s shown in Table 2-1-1. Using an additive rating scheme with equal rates (a common first screen that one might use to assess feature effectiveness), the resulting "score," $\rho + s$, is the same for both features, implying that they are equal in deterrence effectiveness.

However, Equation (2-1) in this chapter allows calculation of the steady-state amount of counterfeit currency in circulation. It is assumed, for convenience,[1] that $\theta = 50$ (that is, average time between passing attempts is approximately 1 week), $\beta = 10$ (average time required for removal of the counterfeit note from circulation is approximately 1 week), and the stockpile confiscation rate c is proportional to the production rate ρ. (This calculation is presented in more detail in Appendix E.) Then Equation (2-1) indicates that Feature A is three times more effective at decreasing the amount of counterfeit currency in circulation than is Feature B. Even though the two features appear similar—even in terms of their overall "score"—Equation (2-1) suggests that they are different in their effect on deterring counterfeiting.

TABLE 2-1-1 Counterfeit Production Rate and Passing Fraction for the Flow Model

Feature	Counterfeit Production Rate ρ	Passing Fraction s	Amount Circulating $y(t = \infty)$
A	0.9	0.1	0.02
B	0.1	0.9	0.06

[1]Compare these assumptions to the *tacit* arbitrary assumption in the additive rating scheme that equal weights should be given to ρ and s, whereas the model provides a framework for estimating the parameters θ and β.

amount of currency in circulation, to the committee's knowledge they have not been used to analyze counterfeit currency flows. As mentioned above, other models, focusing on different flows, variables, and dependences, could also be used.

EVALUATING SECURITY FEATURES WITH THE FLOW MODEL

The goal of any anticounterfeiting effort is to minimize x, the amount of stockpiled counterfeit notes, and y, the amount of circulating counterfeit notes. A flow model can show how these amounts are affected by various flow rates and parameters.[9] Analyzing and understanding the factors that affect the flow provide a basis for understanding the counterfeiting threats and evaluating the outcomes of potential new features. This section summarizes, by relevant flow model variables, the likely effects of various anticounterfeiting efforts that are related to banknote features. As explained in Chapter 4, these considerations were important in the development of a feature assessment methodology and thus were crucial to the selection of the innovative features that are discussed in Chapters 4 and 5.

Decreasing the Counterfeit Production Rate—Parameter ρ

The first opportunity to combat counterfeiting occurs with the deterrence of note production. An array of new, digital tools is increasingly being used by counterfeiters in their attempts to simulate banknotes. The low cost and accessibility of these technologies make them readily available to potential counterfeiters. These new tools continue to challenge the currency designs, especially image-based features.

One deterrence strategy is to prevent production directly, by limiting the availability, the capabilities, the materials, and the use of counterfeiting technology. In fact, this was the first anticounterfeiting strategy implemented by the U.S. government in the 1860s. The distinctive look and feel of intaglio printing could be achieved only by printing processes that were of very limited availability. Likewise, the substrate paper and some of the specialized inks are exclusive to the Bureau of Engraving and Printing. The usefulness of digital imaging tools can be limited both by technology in the printers and by limiting access to the tools themselves. By international agreement, commercial color printers prevent the printing of banknotes—they recognize certain features of various major currencies and refuse to process the image further. Similar technology is implemented in digital image-processing software and in digital scanners. Although these limitations undoubtedly deter casual counterfeiting, they are not foolproof against a determined hacker. Thus, technology stoppers, even when well implemented, are not a panacea, and if

[9]For example, see Appendix E, Table E-1.

TABLE 2-6 Committee's Analysis of the Usefulness of Current Security Features in Deterring Counterfeiting, Evaluated by Class of Counterfeiter

Feature	Primitive	Opportunist	Petty Criminal	Professional Criminal	State-Sponsored
Human perceptible					
Substrate	●	●	●		
Tactility (or feel)	●	●	●		
Watermark	●	●	●		
Security strip	●	●	●		
Intaglio printing	●	●	●		
Offset color blending	●				
Optically variable ink	●	●	●		
Intaglio microprinting	●				
Offset microprinting	●				
Colored threads					
Machine readable					
Paper fluorescence	●	●			
Magnetic ink feature	●	●	●		
Magnetic ink pattern	●	●	●		
Metameric inks	●	●	●		
Digital watermark		●			
Digital Banknote Detection System		●			
Fluorescent security strip	●	●	●		

NOTE: Human-perceptible features are those that can be verified using normal sight and touch. Machine-readable features are those that can be verified through the use of an auxiliary device. "●" indicates some deterrence value; "blank," no significant deterrence value.

they are poorly implemented they could have serious consequences for the operability of the equipment.[10]

A second way to deter counterfeit production is by incorporating features in the note design that are difficult to simulate. Current U.S. banknotes combine numerous features that present varying levels of challenge to counterfeiters in order to provide a layered defense against an array of counterfeiting threats. Although highly qualitative, Table 2-6 provides some insight for optimizing the use of the limited space on the note and avoiding a saturation of features; the table presents an analysis of the usefulness of current security features in deterring counterfeiting, evaluated by class of counterfeiter. The goal of selecting a feature set is to maximize the amount of time, effort, and skill required to create a convincing simulation of the note. This can be accomplished by selecting a complex design, or by focusing on a few, very-difficult-to-simulate features. The trade-off between a complex note

[10]See the discussion in the subsection entitled "Currency Recognition System," pp. 76-78 in National Research Council, 1993, Counterfeit Deterrent Features for the Next-Generation Currency Design, Washington, D.C.: National Academy Press.

and a simple note is also affected by the ease of authentication, which is discussed below.

Although Table 2-6 indicates that the current feature set is not a significant deterrent to the professional criminal or state-sponsored counterfeiter, there is another line of defense. Since both of these classes stockpile and pass counterfeits in large quantities, they are vulnerable to law enforcement operations. The Secret Service has been very effective at shutting down large, professional counterfeiting operations and minimizing their impact, as discussed in the next subsection.

Emptying the Stockpile—Parameter *c*

There is no opportunity for counterfeit notes seized from a stockpile to cause harm, so seizing stockpiled counterfeit notes is an effective way to disrupt the counterfeiting system. A stockpile can be any storage location prior to the passing of the counterfeit; it could be a warehouse, a suitcase, or the pocket of a criminal.

The possession of counterfeit currency is a crime investigated by the Secret Service. To carry out its mission, the Secret Service works with state and local law enforcement agencies, the U.S. Department of the Treasury, and foreign law enforcement agencies to pursue counterfeiters. The committee believes that the expertise and methods employed by the Secret Service are an important component in the successful effort to reduce the counterfeiting rate of U.S. banknotes below that of other major currencies. The decrease in foreign counterfeiting volume between 1995 and 2004 supports this assertion.[11] While there are no statistics to indicate what fraction of stockpiled counterfeits the Secret Service captures, the fact that the value of seized counterfeits (about $100 million in 2004) is vastly greater than that of passed counterfeits (less than $15 million annually) indicates great effectiveness in the disruption of the stockpiling of counterfeit banknotes.

Disrupting the Passing of Counterfeits—Parameter *s*

A counterfeit note does not profit the maker until it is exchanged for value and passed into circulation; if the note cannot be passed, there is no incentive to counterfeit. Therefore, efforts to combat counterfeiting must include strategies to disrupt the passing process. These efforts typically focus on educating the human cash handler to readily identify counterfeit notes and on employing reliable machine authentication.

People who use U.S. currency are the first line of defense against counterfeiting. Because the primary opportunity to pass a counterfeit note is the first point-of-sale,

[11]The particular effectiveness of the Secret Service is attributed to its unique role as a law enforcement unit with an express charter to counter all efforts to counterfeit U.S. currency.

the key people are those who handle currency as a part of their job. Therefore, the first line of defense is the public, which includes the often-inexperienced clerk at the counter or the taxicab driver and other like individuals who are the most likely points-of-sale for the criminal fraternity. In Canada, studies have shown that 78 percent of counterfeits are detected by individuals and businesses.[12] Even the high-quality supernote was first detected not by a machine but by an experienced cash handler who noticed an improper "feel" to the note.[13] Clearly, the most efficient way to prevent the passing of counterfeit notes is to design banknotes that can be authenticated at the point-of-sale, and the optimal human-perceptible security feature is *visible*, *usable*, and *known*.

An effective *visible* feature is one that is easy to authenticate regardless of light levels and the fitness of the note and that therefore creates minimum delay at the point-of-sale. A *usable* feature does not require out-of-the-ordinary methods such as a magnifying glass, transmitted light, or a machine. A feature that is both visible and usable provides an unambiguous signal in a short amount of time in a variety of environments. This feature is not obscured by a busy or distracting design or by "feature overload." Because rejecting a banknote is an undesirable event to most cash handlers, the importance of both speed and certainty in feature evaluation cannot be overstated.

The last factor, ensuring that a feature is *known*, is the most elusive. Knowledge of features does not necessarily mean that a user can name them, but such knowledge may be evinced in an unconscious, habitual awareness. The number of features and the fact that older banknote issues are not devalued can make such education a complex and layered undertaking.

As a result of the public education effort for the new $20 note, public recognition of the currency features increased to 85 percent in the United States. While the impact of feature recognition has not been quantified in the United States, similar educational efforts in Canada have dramatically decreased the acceptance of counterfeit notes by cash handlers.[14] Thus, educating cash handlers to recognize security features is undeniably an important factor in counterfeit deterrence.

One of the most basic yet revolutionary methods intended to disrupt passage of counterfeits is the growth of machine authentication. This is not a new concept—in the 18th and 19th centuries, gentlemen of business carried pocket-size scales to ensure that the coins they received were of the correct weight for their value. In those days of coin shaving, it was prudent and socially acceptable to authenticate coins before accepting them. In the 21st century, the concept is being revisited by electronic currency authenticators.

[12]J.F. Chant. 2004. Bank of Canada Review. Ottawa, Ontario: Bank of Canada. P. 45.
[13]S. Church, Bank of Canada. 2005. Presentation to this committee, July 21.
[14]S. Church, Bank of Canada. 2005. Presentation to this committee, July 21.

Over the past 15 years, advances in low-cost sensor technology have enabled the machine-reader market to grow in size and scope, while machine readers shrink in cost and footprint. Currently, over $64 billion passes through vending-type note acceptors per year in the United States, with 10 million to 20 million daily transactions;[15] in addition, the retail and banking sectors process billions of dollars through machine counters yearly. As sensors continue to decrease in size and cost, machine readers will add authentication capabilities and will become pervasive, particularly in consumer cash-handling applications.

Because notes rejected by a machine reader at the point-of-sale are currently neither tallied nor removed from circulation, no comprehensive statistics are available on the impact of machine readers on the passing or attempted passing of counterfeits. However, industry representatives report that even low-end denominators detect and capture 90 percent of domestic counterfeits, with authenticators nearing 100 percent.[16] Therefore, machine readers represent a significant deterrent to the passing of counterfeits. Because rejected notes are typically returned to the consumer, this tool does not require the cash handler to deal with the consequences of accepting a counterfeit, making machine readers particularly attractive to the retail sector. Such authenticators are already in widespread use in Europe and Canada, and their offer of hassle-free security from counterfeits may be attractive to U.S. retailers as well. An increase in the use of machine authenticators could result in a commensurate decrease in the frequency of counterfeits passing.

As machines replace human cashiers, they may increasingly become the first line of defense against the passing of counterfeit notes. It is therefore important to design notes for reliable machine authentication. Current machine denominators and authenticators sense a variety of features, and it is important for currency design to maintain and add features that target the requirements for both human and machine authentication. Specifically, the ideal machine-readable feature would have these characteristics:

- It could be sensed by a point sensor moving across the note;
- It would be readable from both feed directions—wide-end-first and short-end-first;
- It would emit a strong and reliable signal;
- It would be independent of the orientation and face of the note;

---

[15]R.R. Bernardini. 2004. New security features and their impact on low-cost note readers. Proceedings of SPIE, Optical Security and Counterfeit Deterrence Techniques V, R.L. van Renesse (ed.), Vol. 5310, pp. 52-62.

[16]Staff members, Cummins-Allison Corporation, Mount Prospect, Ill. 2005. Discussions during a subcommittee visit, October 7. It should be noted that these statistics refer to controlled experiments. In field use, counterfeits are just another class of notes rejected by the denominator or authenticator device that may not like perfectly authentic notes for any of a variety of reasons.

- It could be reliably located despite manufacturing and reading tolerances; and
- It would provide series identification.

Removing Counterfeits from Circulation—Parameters r, h, and β

Once a counterfeit note is finally identified, it should be removed from circulation. Even though repeated passing does not result in additional monetary damage, the final person in the chain will feel the impact. Unfortunately, removing counterfeits from circulation is not always straightforward. Several factors affect the efficiency of counterfeit removal and any subsequent law enforcement investigation.

Currently, there is no incentive other than "good citizenship" for the public interception of a counterfeit note close to the initial distribution point. In fact, there is a disincentive to turn in a counterfeit, as the finder receives no compensation and may even be subject to a considerable time penalty from questioning by law enforcement officers.[17] The committee understands that a holder of a counterfeit note on balance might rather destroy it (or keep it as a souvenir) rather than contact the authorities. Similarly, manufacturers of currency-handling equipment report that some retailers tend not to purchase dedicated counterfeit-detection modules for their machine counters in order to avoid the inconvenience of dealing with counterfeits that cash handlers had accepted at the point-of-sale.[18] There is no question that such an attitude toward counterfeit identification interferes with the identification and removal of counterfeit notes, as well as with the eventual capture of the responsible counterfeiters.[19]

It may be possible to provide incentives to citizens who turn in counterfeit notes. A primary objection to such a policy is that it would provide an opportunity to counterfeiters to exchange their products for genuine currency, and it might engender an impression in the public eye that counterfeiting is a victimless crime in which the government sustains the loss. However, a system that reimbursed or even rewarded cash handlers for turning in counterfeit notes when and if the counterfeiter was convicted would overcome some of these objections: certainly there would be no benefit to counterfeiters for turning in their own notes, and recipients of counterfeits would be rewarded for their role in the criminal prosecution. Regardless of the details, it is important that any incentive system decrease

[17]A well-meaning citizen will obviously receive a negative financial impact as well, because the counterfeit must be surrendered to authorities without compensation.

[18]Staff members, Cummins-Allison Corporation, Mount Prospect, Ill. 2005. Discussions during a subcommittee visit, October 7.

[19]The Secret Service estimates that up to five times as much counterfeit currency is in circulation as is recovered each year.

the time, trouble, and stigma of turning in a counterfeit. A culture that encourages finding and turning in counterfeits to authorities would be a valuable aid to law enforcement.

As currency circulates, particularly domestically, it is processed through the banking system, often many times before it wears out. Branch banks typically process their cash deposits through desktop currency-counting machines, which incorporate authentication technology. Nearly all branch banks also use machine counter-sorters that can sort, face, count, authenticate, stack, and band notes. Both of these classes of machine have a near-perfect identification record for counterfeit notes, so when these machines are used, passed counterfeits are removed from circulation as they pass through a bank. Any counterfeit notes that might escape the branch banks' systems are captured at a Federal Reserve Bank, which uses a proprietary machine reader that senses a wide range of overt and machine-readable features.

CONCLUSIONS

- Five classes of counterfeiter—primitive, opportunist, petty criminal, professional criminal, and state-sponsored—can be characterized by the technologies and the methods they use. The counterfeiting threat posed by each of these classes is somewhat different. Opportunistic, casual counterfeiters pose a particularly onerous threat to the integrity of currency, given the diffuse nature of their counterfeiting activities enabled by affordable, highly capable advanced reprographic and digital technology.
- The counterfeiting threat can be described by an analytical flow model that relates the steady-state amount of counterfeit in circulation to a variety of measurable variables.
- Strategies that most effectively deter production, disrupt stockpiling and passing, and remove counterfeits from circulation may differ for each of the five classes of counterfeiter.
- An analytical flow model can provide a quantitative basis for prioritizing features and feature sets so as to provide the best overall defense against counterfeiting.

3

Emerging Counterfeiting Technology Threats

Current digital image capture and reproduction technology has evolved rapidly in recent years, resulting in the affordability of highly capable desktop systems for the home consumer market. Most home users want to take a picture with their digital camera or scan an image with their scanner and then print out a faithful reproduction of the original. While very good images can be readily obtained today, faithful reproductions are not so easy or repeatable; color-management tools are thus available in image-processing software for delivering good color reproduction. A skilled user can match colors easily, and complex image-processing tools are becoming accessible even to casual users. The digital technology revolution has had a concomitant effect on the capabilities accessible to the counterfeiter. As discussed in Chapter 1, this increase in capabilities and access to them provide a major motivation for this study. The charge for the study (see Appendix A) requests the committee to make an assessment of technology threats with respect to U.S. currency. This chapter describes the current state of the art for, and possible future innovation in, relevant image capture and printing. The chapter also discusses the implications of the identified technology threats for U.S. currency.

DIGITAL COUNTERFEITING

The counterfeiting of currency, or indeed other documents, employs each of the three major steps involved in digital imaging: (1) capturing the image, (2) processing the image, and (3) printing the image. These three steps of digital imaging may be carried out in either binary or analog mode, and there are also hybrid processes and steps that can be done in various combinations. Binary and analog images both have advantages and disadvantages. Producing an image in a binary system requires patterning the image in such a way as to make it appear to have gradations. However, on close inspection, this type of pattern is revealed to be dense on-and-off spatial patterns, with each pixel printed with either ink or no ink. Analog printing produces images with an array of elements that have different optical densities. Hybrid systems draw from both paradigms, but the results tend toward the artistic rather than the production of multiple images that are all closely identical. Images that appear on Federal Reserve notes (FRNs) can be acquired using a variety of digital scanning and photographic equipment—complemented with sophisticated art software—that offers very cost-effective capture of images with good quality. Table 3-1 summarizes the committee's analysis of the usefulness of security features in deterring counterfeiting that uses digital age tools.

The committee anticipates that the capabilities of digital imaging will improve over the next 5 years. This chapter describes the expected improvements in digital imaging technology and how these improvements are likely, in the committee's opinion, to affect the capabilities of the classes of counterfeiter described in Chapter 2.

TABLE 3-1 Summary of the Committee's Analysis of the Usefulness of Current (2006) Security Features with Significant Deterring Value in the Production Rate of Digital Age Counterfeiting

Features	Ink-Jet Printer	All-in-One Device	Color Copier	Flatbed Ink-Jet Printer	Digital Press	High-Quality Scanner	Imaging Software
Human perceptible							
Substrate	●	●	●	●	●	NA	NA
Tactility (or feel)	●	●	●	●		NA	NA
Watermark	●	●	●	●		●	●
Security strip	●	●	●	●		●	●
Intaglio printing	●	●	●	●		NA	NA
Offset color blending							
Optically variable ink	●	●	●	●		●	●
Intaglio microprinting		●	●				
Offset microprinting		●	●				
Colored threads							
Machine readable							
Paper fluorescence	●	●	●	●	●	NA	NA
Magnetic ink pattern	●	●	●	●		NA	NA
Infrared ink pattern	●	●	●	●		NA	NA

NOTE: "●" indicates some deterrence value; "blank," low or no deterrence value; "NA," not applicable.

IMAGE ACQUISITION

To replicate an FRN using digital technology, a counterfeiter must first transform an analog object, the physical note, into a digital format. U.S. currency is approximately 2.5 by 6.0 inches. Therefore, the scanned image of a U.S. banknote from a 3,000 pixel per inch (ppi) scanner would require about 7,500 by 18,000 pixels, or a total of 135 million pixels. This file size is now well within the capabilities of even lower-priced home copiers and amateur flatbed scanners. Table 3-2 provides a summary of the committee's analysis of the capabilities of image-capture systems and relative costs.

The majority of the printed features on FRNs are binary, meaning that at every pixel element, ink is either deposited on the substrate or it is not. Virtually all of today's high-quality printing processes are also binary.[1] However, each pixel of an image from an input scanner is usually recorded with at least 8 bits, or 256 levels of intensity. When the image is prepared for printing, however, the bit depth is reduced again to 1 bit because most electronic printers are binary devices. Most often, because halftoning or similar techniques are used to generate the appearance of gray in the image that is printed, the sharp binary image is lost in the transition to analog and back again.

Flatbed input scanners costing a few hundred dollars or less can now scan at high quality, at reasonable speed, and with spatial pixel densities up to 43,000 ppi using a single silicon sensor chip that has sufficient sensors to cover the entire page image—resolution of 3,000 ppi requires about 25,000 sensors to cover an image 8½ inches wide. This resolution is more than adequate to capture most printed material very well, especially with some postprocessing of the image to enhance fine features. Increasing the scanning density beyond current capabilities is unlikely, because improved scanning pixel density is not needed for most applications and because the commercial profitability of flatbed scanners is not high enough to warrant large investments for specialty or niche applications—increasing this density would require costly improvements in optics and data storage.

A scanning resolution of up to 4,000 to 5,000 ppi is currently available on the best graphic arts scanners. These scanners are often referred to as copy-dot scanners because they attempt to copy all of the halftone dots in the original. Thus, advances in image capture in such devices will not improve image acquisition for counterfeiting because the resolution is at a practical limit already.

Today, digital cameras cannot provide the image quality of a flatbed scanner owing to many optical effects as well as to limited pixel density. But digital cameras, including those in cellular telephones, will certainly continue to improve, and

[1]Photographic film and thermal dye printers, being analog rather than binary technology, are two exceptions.

TABLE 3-2 Summary of the Committee's Analysis of the Capabilities of Digital-Image Capture Systems

Capture Device	Resolution (pixels per inch)	Cost	Comments
Digital camera	Varies	Low to high	Limited number of total pixels available
Simple consumer-grade reflective flatbed scanner	2,500	Low	Ideal for counterfeiting owing to high quality and low cost
High-quality reflective flatbed scanner	2,500-4,000	Moderate	High quality with only moderate cost
Professional drum scanner	4,000-5,000	High	High cost presents barrier to entry
Artwork software	No limit	Free to moderate	Time-consuming and limited to use by experienced artists

performance-for-cost will significantly increase in the next few years. Nevertheless, these improvements are not likely to provide the counterfeiter with any additional or lower-cost capabilities than those currently available from other technologies.

IMAGE PROCESSING

Once an image has been captured, it must undergo a number of processing steps before the best possible reproduction can be made. The steps are these:

- Removing scanning artifacts such as dust specks;
- Adjusting the brightness and contrast of the image;
- Performing color adjustments;
- Applying software filters to enhance edges and sharpen the image, often called unsharp masking; and
- Rotating the image if it has been scanned at a small angle.

Digital image processing of FRN images can be done using a wide variety of software tools, ranging in cost from free to very expensive. Achieving expert results with such software is at the present time highly dependent on the skill of the user.

Current Capabilities

One of the significant improvements in image processing has come about through generally available software. Packages such as Adobe Photoshop, Microsoft's Digital Image Pro, and others provide powerful tools for improving and enhancing digital images. Personal computer (PC) operating systems also are sold with image tools that can be accessed by the most casual user. The capabilities of the very-low-cost tools are expected to increase with new operating system releases. The growing demand for accessible digital photography tools makes image processing a very rich area of innovation that will be available to the counterfeiter. Additional tools available through online photo-sharing Web sites can easily—and anonymously—improve the reproduction of color images.

An option for any counterfeiter would be to inspect the currency visually and then to re-create it line by line using software illustration programs; this would be a painfully slow process, even though it would allow great detail to be reproduced well. Bitmap editing, or pixel-by-pixel editing, of a captured image can be done using very inexpensive software such as Microsoft Paint, but this too would take a tremendous investment in time and a good dose of eyestrain to generate a useful image. The labor content of such an approach is not expected to abate in the next 5 to 10 years.

An alternate approach is to re-create an FRN feature by feature; programs such as Adobe Illustrator, Adobe Photoshop, Adobe PageMaker, QuarkXpress, and others provide a rich set of tools with which to generate exceptionally high-quality images. In addition, these programs can leave the created or re-created data in a form that allows "resolution-independent" printing options—meaning that the quality of the output is materially dependent on the quality of the printer. Using an illustration program in this manner is analogous to engraving a false duplicate by hand.

The scanning of sharp lines such as those found in intaglio printing is a particular challenge. Often, the images of lines scanned at angles or lines that are wavy or that possess other artistic features will not necessarily reproduce with uniform thickness in an image. When printed, the lines can vary by a pixel or so, which can be quite visibly apparent under certain circumstances. Often such lines need to be retouched to remove problematic artifacts. Software to perform this task automatically is under development and has the potential to greatly reduce the labor cost of making high-quality images and to eliminate the need for special skills and additional steps.

To prepare an image for most digital printing, the scanned image needs to be converted to a binary format. The scanned image can be halftoned,[2] as in conven-

[2]*Halftoning* is the transformation of a grayscale or color image to a pattern of small dots with a limited number of colors—that is, just black dots for grayscale images or a combination of cyan, magenta, yellow, and black dots for color images—in order to make the image printable. In the basic case of gray scale, the halftone process creates patterns of small black dots on a white background. When viewing such an image from a sufficient distance, the human viewer cannot see the dots themselves because they are too small. Instead, the human viewer has the illusion of seeing gray, whose

tional printing, or it can be made without halftone patterns using an on-off process often referred to as thresholding. While technologies are in development, it seems that the practicalities of binary printing processes will be limiting for some time.

The use of masking software, automated subroutines, and other such software shortcuts can greatly reduce the amount of work required of a determined counterfeiter who uses current image-acquisition technology. Artistic design software is improving to the extent that each generation of upgrades makes it easier for the skilled user to reproduce complex effects in the scanned image. In addition, software that seeks to reproduce the look of traditional art materials has advanced by leaps and bounds in the past few years and continues to improve. These developments may threaten one of the most counterfeit-deterring features of U.S. currency—the feel of the intaglio printing.

Emerging Capabilities

Image processing will no doubt be an area of significant development in the next 5 or 10 years. The modern PC can now readily handle gigabyte-size images, more than enough storage for the captured image of an FRN scanned at 4,000 pixels per inch. Because capture quality is already at a high level of fidelity, image processing can be used to enhance the scanned image and can enable high-quality printing. However, most of the software available today, while having somewhat sophisticated automated tools, delivers the best results when it is in the hands of experienced users.

In the next 5 years, much of what occurs today in application software may be incorporated in the scanner or operating system. The growth of the digital imaging market will require that such capabilities be close to automatic. A few years ago, one could do a very good job of processing color film and prints at home. To do so required a darkroom, chemistry, some basic photoinstrumentation, and a lot of experience if it was to be done well. As is clear from any number of trends, the darkroom is now digital, and today's experienced digital imaging enthusiasts are constantly improving their skills. However, like their predecessors in chemical photography, most do not "mix their own chemicals"; they use what is commercially available. Several products introduced in 2005 will automatically produce multiple versions of a color-corrected image, requiring the user simply to select the best image from a group.

---

darkness depends on the amount of black dots on the white background. In color halftoning, the cyan separation, the magenta separation, the yellow separation, and the black separation combine in a set of patterns printed over each other, which the human viewer from a distance observes as a color image.

Technology development will result in increasing automation in image processing in the next 5 to 10 years. Techniques such as automatic contrast adjustment, unsharp masking, line-width control, and feature smoothing are all under development as automatic features. The driver for these changes is to make the user's experience as pleasant and simple as possible. Automatic comparison of the input and output image is currently available in some software packages. Significant improvement is expected in image reproduction because the market will demand it.

In 1981, SciTex provided a stunning image-processing system for graphic artists that cost approximately $1 million. Today, systems that dwarf SciTex system capabilities can be purchased along with their host PC for less than $2,000, a 500-fold performance-for-cost improvement. There is no indication that we are nearing the limits of software or computing technology. The ability to use image-processing software to customize individual notes by changing the serial numbers, plate numbers, or other identifiers may enhance the ability to pass a counterfeit note.

IMAGE PRINTING

Image printing is the final step in digital reproduction except for finishing operations, which would include cutting, trimming, and the addition of simulated security elements. It is also the limiting step in the counterfeiting process. Modern digital printers (see Table 3-3) available to ordinary users can provide quality at or above that available from photochemistry just a few years ago. Some of the improvements that might be expected in printing technology in the near term are discussed below.

Electrophotography

The electrophotographic process is the basis of the most widely used document-copying machines.[3] It begins with a photoconducting surface that is uniformly statically charged. In many copiers, this is a metal plate with a selenium-based coating. The charged surface is then exposed to a pattern of focused light, usually from a laser or light-emitting diode. This pattern is the image to be printed. Where light falls, the charge dissipates and a "charge image" of the light pattern remains on the photoconductor surface.

The image is developed by dusting the charged surface with a pigmented powder, called toner, which is attracted to the charged areas of the pattern. The powder is then transferred electrostatically to paper and, finally, the toner is fused to the

[3]Xerox Corporation was the principal investor in electrophotography in the early days of this technology, which became more commonly known as "xerography," apparently a catchier name than electrophotography.

TABLE 3-3 Summary of the Committee's Analysis of Some Printer Technologies and Capabilities

Technology	Resolution (μm)	Cost	Comments
Electrophotographic (laser)	7-10	Low	Potential primarily for low-volume counterfeiting
Ink jet	7	Low	Attractive for all classes of counterfeiters owing to low cost and high resolution
Thermal	2	Low	Limited by low-image-quality substrate chemistry requirements
Chemical	NA	High	High cost, variable quality, and limited availability of adequate substrate materials
Photographic	NA	High	Produces quality similar to that obtainable with ink-jet printers but at a higher cost

NOTE: For the cost of equipment acquisition, "low" refers to the cost for a typical peripheral for a home computer and "high" is outside this range; "NA," not applicable.

paper with heat. A continuously rotating metal drum moves the plate and paper through the various steps—charging, exposing, developing, and transferring—in a seamless manner.

The quality of electrophotographic images has improved continually since the introduction of this technology in the mid-1900s. The critical elements of electrophotographic technology for counterfeiting are the development, transfer, and fusing steps that control the image quality. Advances in these capabilities have resulted in higher quality and lower cost.

In the development of the image, in which the toner material forms the text and images on the printed paper, the size of the toner particles controls the print resolution. As first implemented, particle size in toner averaged about 12 micrometers (μm). In order to improve image resolution to 600 dots per inch, the particle dimension was reduced to 8 μm and the uniformity of particles was improved. Further reductions in size will be necessary for improvements in resolution. Toners that are substantially smaller than this are not only hard to manufacture properly but also are difficult to keep confined to the printer itself—that is, they can be become airborne. Toners with particles as small as 1 or 2 μm can be made, but they can be hazardous if they become airborne and are inhaled by users.

Advances in the materials properties of toner can also help improve resolution during fusing and transfer. By varying the chemistry of the polymer in the toner, printed images can be made to look more like ink on paper rather than having the look of electronic printing. The quality of the fusing techniques can have serious implications for durability. The fusing techniques of lower-cost printers often are not as durable as those of the faster, more-expensive machines. Folding or creasing of the print can cause poorly adhered toner to flake off at the crease, thus exposing the white substrate and producing an obvious defect. As fusing technology improves, home printers could have the potential to do one-pass, two-sided printing. This means that the registration of the images on the two sides of the substrate can be significantly better than that achieved by removing the printed sheet, flipping it over, and reinserting it for printing of the second image.

A variation of electrophotographic printing is liquid electrophotographic (LEP) technology. Digital printing presses using this LEP technology have micron-sized toner particles, enabling resolutions finer than those typically possible with dry-toner devices. LEP devices are being developed to rival offset presses for applications such as marketing brochures and photograph albums and so are advancing the limits of color reproduction and line art. Existing products do not typically have provisions for highly accurate front-and-back registration.

While the quality of electrophotographic products is expected to improve in the next several years, significant engineering challenges remain that will limit the performance of these systems. Improvements in ink-jet printing, the major market competitor for lower-cost electrophotographic printers, will, however, continue to drive print quality.

Ink-Jet Printing

While ink-jet printers predominate in the home printer market, they are also capable of producing some of the highest-quality images of any type of printer. An ink-jet printer is any printer that shoots extremely small droplets of ink onto paper to create an image. The droplets form small dots on the printed page. These dots can range in density from 10 to 30 dots per millimeter.

The placement of the dots can be very precise. Ink-jet technology is particularly useful because registration in the system is an electromechanical issue that can be solved by straightforward engineering methods and techniques. For this reason, flatbed ink-jet printing has the greatest potential for future use in home printing of high-registration, simultaneously printed front-and-back images.

One disadvantage of this method is ink bleeding into the paper, resulting in a blurry appearance on some types of paper. These effects may be pronounced with certain types of paper, especially when the ink is water-soluble. The introduction of ultraviolet-curable, water-impervious ink-jet inks would solve the problems of stability, lightfastness, water solubility, dot gain, and spread.

Ink-jet printers fall into three categories: continuous, thermal, and piezoelectric ink jet. Their features and applicability to counterfeiting are described next.

Continuous Ink Jet

Continuous ink jet is one of the oldest ink-jet technologies in use. It is fairly mature and arguably produces the best-quality image obtainable on any type of printer. Continuous ink jet is actually better than photography in many cases in reducing image noise. Continuous ink-jet printers are not, in general, printers for home use. They are often large and require a fair level of maintenance and operator skill to attain top performance from the device.

In continuous ink-jet technology, a high-pressure pump directs liquid ink from a reservoir through a microscopic nozzle, creating a continuous stream of ink droplets. One of the advantages of continuous ink-jet printing is the very high velocity of the ink droplets, which allows the ink drops to be thrown a long distance to the target. Another advantage is freedom from nozzle clogging, as the jet is always in use. Another key advantage of continuous ink-jet printers has been their ability to produce small drops of ink that can be repeatedly printed at the same or carefully adjacent spot to vary the effective droplet size. This capability produces stunning quality.

It is not expected that continuous ink-jet printer technology will move significantly forward in use in the next 5 to 10 years for two reasons. First, they are inherently complex to maintain and operate. Second, newer ink-jet printers or other types are now approaching the same quality levels attainable using continuous ink-jet technology.

Thermal Ink Jet

Thermal ink-jet printers, also known as bubble-jet printers, are widely available for home use. These printers operate by rapidly heating a small volume of liquid ink and forcing a steam bubble to form that ejects the ink from an orifice. As the bubble cools, the vacuum created draws fresh ink back into the nozzle. A large investment in this technology has produced a remarkably reliable process. In 2006, the print cartridge is a highly engineered device with 64 or 128 tiny nozzles that are produced using photolithography. To produce an image, the printer runs a pulse of current through the heating elements. Each of the tiny chambers can fire a droplet simultaneously to produce the image.

A disadvantage of this technology for counterfeiters is that the ink must be water-based. Such inks are inexpensive to manufacture, but they may perform best on specially coated media that do not have the feel of currency paper. However, because the print head may be produced at less cost than that for other ink-jet technologies, it is a relatively low-cost process.

One area of active research and engineering is currently aimed at producing multiple drop sizes on demand. However, it is not an easy task to control the bubble and its associated heat cycle to produce different-size bubbles reliably. Possible advances in the next 5 or 10 years may also lead to increased process speed. It is not clear at this stage if either of these advances is likely to be realized for home users or only for higher-end users.

Piezoelectric Ink Jet

Most higher-end ink-jet printers use a piezoelectric crystal in each nozzle instead of a thermal heating element. "Piezo-jet" printers utilize a piezoelectric actuator to produce pressure in the liquid-ink chamber. The pressure and drop size generated by the piezoelectric actuator can be controlled better than in bubble-jet printers and can also produce smaller droplets of ink. These variable-size drops and smaller droplets can generate impressive images. Today, the smallest commercial droplet size available is in the range of 1 picoliter. This corresponds to a drop diameter of about 7.1 μm, or about 3,500 or so drops per inch. Such small droplet sizes are generally not attainable in the bubble-jet schema. However, this very small drop size can be lost on typical paper used for home printing, so it is unknown whether the investment in such improvements would be profitable in the home market over the next 5 to 10 years.

An advantage of piezoelectric ink-jet printing is that it allows a wider variety of inks than does thermal or continuous ink-jet printing. The emerging ink-jet material-deposition market uses ink-jet technologies, typically piezoelectric ink-jet, to deposit materials on substrates. Printers that can deposit glues, resins, or waxes onto a variety of substrates have been available since the early 1990s; they are called solid ink-jet or wax-jet printers. While it has advantages in color intensity, solid ink-jet technology has limited droplet size compared with that of bubble-jet or piezo-jet liquid-ink printers because the viscosity of the melted material complicates the jetting of small drops.

The piezoelectric ink-jet technology is currently available on commercial flatbed digital presses, and while they are currently very expensive, they are easily and quickly configured for counterfeiting. This is a major departure from the setup time required for commercial printing presses; it could allow counterfeiters who might work in printing shops to use the digital equipment without alerting their management. In addition, advances in this technology in the next 5 to 10 years may mean that this level of quality may be available for consumer units. Advances may also mean that metallic or plastic particles could be "printed" and someday may be able to simulate some non-image features. For these reasons, piezoelectric ink-jet printers may prove to be the most useful technology to counterfeiters now and in the future.

Thermal Printing

Thermal printing commonly is done in one of three forms: thermal transfer, thermal dye transfer printing, and custom-substrate thermal printing. Thermal transfer, one of the early printing technologies, has come and gone as a technology of choice. Its main advantage has been its reliability and high edge acutance. However, achieving printer pixel density higher than 1,000 ppi is problematic owing to the requirements for the heater heads. It is not expected that thermal transfer will be able to compete with bubble-jet, piezo-jet, or electrophotographic images as time goes on.

The second form of thermal printing, thermal dye printing, has also been referred to as dye sublimation printing. However, in most cases, sublimation is not the physical process that causes image formation, and hence "thermal dye printing" is the preferred nomenclature. Thermal dye is much like thermal transfer, except that instead of transferring material from a ribbon surface to a substrate, dye is migrated from a ribbon and absorbed into a special substrate under pressure. The main advantage of thermal dye is that it can produce true grayscale images without any of the usual halftone patterning or other geometrical structures that trick the eye into seeing gray scale. Such printers produce images very close to those achieved in photography. The main disadvantages of thermal dye are limitations in spatial resolution and speed. Currently, about 600 ppi is the maximum practical density, which is generally low for quality counterfeit currency generation.

The third form of thermal printing uses custom chemically treated paper, and it prints with heater heads not unlike those used in thermal transfer. It generally cannot produce good colors, and like the other forms of thermal printing, this technology is generally low quality and is unlikely to improve in the future to any extent. The specialty substrate, in which the chemicals used for image generation are in the paper, would significantly complicate its use for currency counterfeiting.

All of these technologies are slow, expensive, and require special substrates to achieve the promised quality. Because none of them is a focus of current industry innovation, concerns about future thermal printing developments with respect to counterfeiting are likely unwarranted.

Chemical Printing

No imaging process has seen its world grow smaller more quickly than that of conventional chemical photography. Photography can produce outstanding images, but such images are no longer unique to this process. Electrophotographic and ink-jet processes can produce images in many ways as good as those of photography. Because genuine currency is produced in general with commercial processes—offset, intaglio, and letterpress printing—the images are binary, meaning that each pixel on the banknote either has ink or it has no ink. Hence, photography offers little value over simpler digital processes that are much more reliable and lower in cost. Photography is still of use in the generation of printing plates, but this area is also quickly becoming digital.

There is little current activity in either using or improving the quality of nonphotographic chemical printing, and it is not expected that this technology will benefit counterfeiters in the foreseeable future. Problems of chemistry control, quality, and maintenance were the factors that killed the early chemical copiers when electrophotographic processes came on the market. These factors are still a barrier to this technology today.

Image-Printing Implications

Each type of digital printing method discussed above has its own special substrate requirements for optimal image printing. While new substrate capabilities are always emerging, the nature of the various electronic printing technologies restricts the range and types of substrates that work well. These issues are discussed in Box 3-1.

The various electronic and other printing processes described above have varied advantages in image quality. The quality of digital presses and advanced electrophotographic printing has risen to the point that it is expected to replace offset printing in the near future. This means that the value of using offset printing as a tool to deter the counterfeiting of the images on banknotes is limited.

As is image capture, image printing is now at pixel density levels—2,400 to 3,600 ppi—that are at the limits of human vision. However, improvements in image quality could come from printing images other than in binary mode. Current ink-on-paper printing such as offset and intaglio are binary processes. To simulate the look of gray—as in a watermark—commercial printing devices and most electronic printers print black ink in a patterned fashion. This patterning results in "aliasing" artifacts, because lines are not single strokes but a collection of dots that look at normal viewing distance like single strokes.

New approaches in both ink-jet and electrophotographic printers can produce some levels of gray, and multiple gray-level printing could produce dramatic improvements in image quality. In electrophotography, toner particulate size control and toner management as well as fusing technology (which depends on toner chemistry) will continue to set the limits for electrophotography laser printers.

Current ink-jet printers must produce only one drop of ink at a time, and each drop must be placed adjacent to the previous drop or drops to produce a line. Such lines often look a bit ragged and do not reproduce intaglio-printed structures well. The size of the ink-jet orifice and the ability to print reliably using small orifices will set the limits for ink-jet pixel density. Inks with multiple densities of the same hue

BOX 3-1
Limitations on Electronic Printing Technologies Regarding Substrates and Additional Elements

Most types of paper work well in electrophotographic printers. The two principal restrictions in these printers are the ability of the paper-feeding system to handle the substrate material and the substrate interactions with the electrostatic subsystems such as image transfer and fusing. Many electrophotographic printers can easily handle light- to heavyweight papers, well within the range of currency substrates. Some newer printers can print on both sides automatically, but back-to-front registration of printed features would in general not be nearly as good as that achieved in offset printing.

The look and feel of the substrate—including the paper surface, stiffness, and intaglio texture—are difficult for most counterfeiters and technologies to reproduce using current desktop image-reproduction technology.

Ink-jet printing can be either more or less tolerant of substrate differences, whereas electrophotographic printers are less sensitive to substrate quality. Paper-feeding requirements are still important, but ink jets can generally feed a wider array of substrate thicknesses. Card stock, shiny or matte finishes, and vellums can be used, because the paper path is usually very short and throughput speeds are low. The only additional requirement of substrates used with ink-jet printers is that they permit the ink to dry and adhere to their surface properly. Liquid ink must to some extent wick into the substrate surface for adhesion. Solid ink is almost substrate-independent, although the quality of the image might vary.

Features that affect light transmitted through the substrate—the watermark and security strip—also pose challenges, because most image reproduction is based on reflected light. Banknote features used in machine authentication (e.g., special inks) are difficult to simulate with desktop printing technology.

While thermal printers are unlikely to be used for counterfeiting, they may be used as part of a larger system. For example, laser thermal printing could be used to apply special symbols or thin foil features over a previously ink-jet-printed page. Electrophotographic printing can also be used to affix foil or holographic features.

may achieve different tones in the image but with much-reduced patterning. This approach has the advantage of using existing transport and mechanical systems and changing only the supply packages for printers in place.

Thermal printers can produce excellent vertical or horizontal lines, but their limited pixel density compared with that of electrophotographic and ink-jet printers is expected to render them inferior both now and in the future.

Market drivers are a final consideration for high-end printing equipment. One result of gains in desktop technologies and their capability to produce professional-looking results is that small professional printing houses across the country are going out of business and often selling their equipment for as little as 10 percent of its original cost. This may bring a professional setup well within the price range of the petty criminal. With a good computer, a pirated copy of Adobe Photoshop, a digital direct-to-plate machine, and a used Heidelberg press, a petty criminal with considerable skill might start an offset money-printing workshop in his or her garage. Straight-to-plate technology, in which a computer design can be digitally inked onto printing plates, can turn older equipment into very modern presses. This technology in particular may be a threat because the ease of setup is a significant advantage.

Emerging Technologies

In addition to the improvements that are expected in current printing technologies, the emergence of some technologies that go beyond image printing may soon overtake those recognized methods.

The Convergence of Printing, Manufacturing, and Biology

A number of emerging tools use technology invented for printing on paper, but redesigned for other processes. Examples include the high-volume manufacture of microelectronic circuits,[4] smaller-volume prototyping processes,[5,6] and even printing of biological material to aid analysis.[7] Printing-based manufacturing systems offer levels of resolution and registration that significantly exceed the current or projected capabilities of conventional ink-on-paper printers. The target applications for this technology—such as printing flexible electronic circuits—demand very low-cost high-throughput operation, and large-area printing. Goals include rates of 100 square meters per hour, with the ability to process substrates larger than 5,000 square meters with accurate registration.

The range of "inks" that can be printed by these systems includes not only the dyes and toners used in conventional printers, but also color-shifting and other optically active inks. Functional materials for the semiconductor elements of thin-film transistors or the electroluminescent layers of emissive displays have also been printed. New ink-jet printing processes can also print "inks" made of a variety of metals, glasses, and plastics, which implies that simple printing processes may be able to simulate non-image features, such as the metallic print on the security strip.

Although access to these manufacturing systems is not expected to be widespread, it will increase if their use broadens to more distributed manufacturing

[4]Proceedings of the IEEE [Institute of Electrical and Electronics Engineers]. 2005. Special issue on Flexible Electronics Technology. Vol. 93.

[5]A.V. Kumar and A. Dutta. 2003. Investigation of an electrophotography based rapid prototyping technology. Rapid Prototyping Journal 9:95-103.

[6]See <http://web.mit.edu/tdp/www/whatis3dp.html>. Accessed February 2007.

[7]See <http://www.shimadzu-biotech.net/pages/news/1/press_releases/2004_07_23_chip.php>. Accessed February 2007.

systems. It is also possible that some of the features of these advanced systems could migrate into low-cost desktop printers.

Some specific technologies in this area include the following:

- *High-speed ink-jet printing* in manufacturing systems uses piezoelectric or thermal print heads for high-speed, large-area patterning—up to 100 square meters per hour. These printers involve hundreds of active nozzles, each operating independently, at frequencies of up to tens of kilohertz. Recent work focuses on applications in the manufacturing of display systems and certain components of packages for microelectronics. Large-area, high-resolution displays that use organic light-emitting diodes or circuits patterned by ink jet are possible. Sophisticated control devices enable excellent repeatability in the printing. Machine vision provides the ability to perform registration in real time during printing. Experimental printers are capable of printing droplets with diameters of 5 microns or less, with registration at even finer scales, particularly when patterns of wetting and nonwetting regions on the substrate are used to confine the printed droplets. The range of inks that can be printed is broad.
- *Screen printing* uses a squeegee-type blade to push viscous inks through patterned openings in a screen mesh. Existing applications include the low-resolution patterning of decals, signs, and textiles. Screen printing is also commonly used to define some features on printed circuit boards. Recent developments indicate that improved printers and screens can achieve resolution near 10 to 20 μm with good registration.
- *Laser-induced thermal transfer printing* uses a focused laser beam to selectively transfer layers of solid-material "inks" from a "donor" sheet to a target substrate. The basic printing mechanism is similar to that of a conventional thermal transfer printer. The use of lasers in place of resistive heaters, however, can improve the resolution significantly, to levels that are comparable to the spot size of the laser, near 2 μm. The overlay registration is as good as the resolution. Large areas and high patterning speeds are possible with rigid and flexible substrates. Inks range from electroluminescent organics, to metals, to colored and black dyes, to semiconducting polymers, to carbon nanotube composites, to biological tissues. The most well-developed potential applications of this printing technique are for the production of organic light-emitting display devices, and color filters for liquid-crystal displays. In the latter case, the printed material consists of stacks of charge transport and emissive layers. In the former, dye-doped polymers are used.
- *Ink-jet printers* can print a wide range of materials in addition to those commonly used for printing images. The "inks" include liquid suspensions of nanomaterials, such as carbon nanotubes and buckyballs, colloidal particles,

nanoparticles, and nanowires—metals and semiconductors, with magnetic and dielectric properties. Biomaterials, including deoxyribonucleic acid, or DNA, have also been printed. Active materials—semiconducting or light-emitting materials—as well as passive dielectric and photocurable polymers are also printable. Many of these unconventional inks might be used to simulate features, such as optically variable images, that appear on currency. Ink-jet technology can also be used to pattern classes of materials that are themselves not directly printable. In these cases, printed polymers or waxes may be used as sacrificial masking layers for patterning other materials. After use, these layers can be removed. In this way, both positive and negative patterns can be produced by ink-jet techniques.

IMPLICATIONS OF TECHNOLOGY TRENDS

In summary, and referring back to the classes of counterfeiter identified in Chapter 2, primitive counterfeiters do not use the types of digital technology discussed in this chapter, but create counterfeits using little more than manual artistry to modify a piece of currency in order to increase its value and obtain financial gain. Opportunist counterfeiters counterfeit occasionally and use typical desktop computer equipment and available crafting supplies, sometimes in creative ways. Petty criminals counterfeit in a dedicated manner and actively invest in specialized computer equipment and materials. Professional criminal counterfeiters focus the efforts of a large group of people on the sophisticated production and distribution of counterfeits. State-sponsored counterfeiters may use the very same high-precision equipment that the government uses to manufacture banknotes. Table 3-4 summarizes the committee's analysis of the digital technology access of the four classes of counterfeiter that use digital technology.

TABLE 3-4 Summary of the Committee's Analysis of Digital Technology Access, by Class of Counterfeiter

Class	Ink-Jet Printer	All-in-One Device	Color Copier	Flatbed Ink-Jet Printer	Digital Press	High-Quality Scanner	Imaging Software
Opportunist	●	●				●	●
Petty criminal	●	●	●			●	●
Professional criminal	●	●	●	●	●	●	●
State-sponsored	Not applicable—reproduces government processes directly.						

NOTE: The primitive class is omitted from this table as it does not use digital technology. Within each counterfeiter class in the table, additional nondigital techniques may be used to improve note simulations—for example, craft supplies to reproduce features that use color-shifting ink. "●" indicates high likelihood of access; "blank," low likelihood of access.

The committee has shown that innovation and skilled engineering will continue to improve the range of excellent, reliable, and cost-effective image capture and printing technology for consumer use over and above what is already available at very affordable prices—although some physical limits may dominate the possible improvements in image quality. As the cost of imaging equipment goes down and print quality goes up, the use of this type of equipment by the opportunist counterfeiter will expand. The same equipment will enable expanded operations by petty criminals, and it may make counterfeiting more lucrative for professional criminals as well. The trend means that the protection against counterfeiting afforded by a two-dimensional printed image casually viewed in reflected light is highly diminished.

Therefore, the committee concludes that image-capture, processing, and reproduction technologies, both current and predicted, pose a significant threat to the security of FRNs. Emerging technologies will continue to limit the ability of any two-dimensional printed image to deter widespread counterfeiting successfully. Images involving other classes of features—images viewed in transmitted light, light-reflecting features, or other complex optical features—offer a substantial challenge to primitive and opportunist counterfeiters and a costly barrier to petty and professional criminal counterfeiters.

But there is also another technology threat, not related to the hardware of image capture and printing but nevertheless a significant, pressing, and perhaps more insidious threat—that is, the augmentation of the counterfeiter's skills owing to improved communication available via the Internet. Counterfeiters today can easily search online for raw materials and surplus high-quality printing equipment. This search capability, coupled with the ability to purchase these materials and equipment from global sources via the Internet, accounts for an important and growing threat.

Information itself is also a precious commodity for the counterfeiter. The information shared around the globe today may include ideas for simulating currency features, novel concepts for combining processes to create a better counterfeit, or expertly processed image files. Successful information sharing may also create new distribution networks for counterfeits. Such network-coordinated distribution would require law enforcement to be at least equally well networked in order to discover and stop it.

CONCLUSIONS

- The Internet remains a growing threat to the security of U.S. currency owing to the ability it affords counterfeiters to augment their knowledge base of how to simulate and reproduce the look and feel of Federal Reserve notes.

- Image-capture, -processing, and -reproduction technologies, both current and predicted, pose a significant known threat to the security of Federal Reserve notes. The security of FRNs often depends solely on the casual viewing of two-dimensional printed features in reflected light, notwithstanding the deterrence value of the feel of the note, security strip, and watermark.
- Future innovations in flatbed scanners and other methods for capturing images are not likely to be an area in which technology will further aid the counterfeiter in the next 5 or 10 years. Scanning quality along with the capture of color features and other image characteristics is already adequate for many counterfeiting activities.
- Printer technologies, including low-cost thermal ink-jet printers for home use and other ink-jet printers and electrophotographic printers, produce counterfeits that can be passed, even though the notes are poor reproductions. However, piezoelectric ink-jet printers may prove to be quite useful to counterfeiters as they provide a noticeable improvement in print quality.
- Improvements in consumer-grade scanners may occur. These new scanners are likely to further enable the opportunist counterfeiter to close the quality gap with today's professional criminal counterfeiters.
- In the future, substantial improvements in the automation of image-processing algorithms are expected to help the less-skilled user process images like a graphics expert. Automated capabilities such as line-width control, uniform image appearance, and color balance would enable an ordinary user to easily obtain an optical image that is very faithful to the original.
- Emerging technologies are targeted at improvements in desktop capabilities; these improvements will continue to limit the ability of any two-dimensional printed image to deter widespread counterfeiting successfully.
- The convergence of printing, manufacturing, and advanced materials technology may offer significant new capabilities to professional counterfeiters in the future.

4

Innovative Counterfeit-Deterrent Features

This chapter summarizes the most outstanding proposed counterfeit-deterrent features that the committee estimates can be developed for implementation in future series of Federal Reserve notes (FRNs) within what is considered to be an intermediate development time of less than 7 years. The chapter also discusses the process used by the committee to assess the value of proposed features and to rank a subset of them for inclusion in this report. The next chapter summarizes the longer-term, disruptive feature platforms.

FEATURE ASSESSMENT

The committee was tasked to "identify features, materials, and technologies to deter counterfeiting of FRNs, and assess their relative effectiveness" against both current and emerging threats. In particular, the committee was asked to consider novel, noncommercial, and potentially unique features that could be incorporated in notes in the longer term and, furthermore, to consider features that would be aesthetically pleasing and amenable to visual authentication. As a secondary task, the committee was asked to evaluate features for use by the blind, for law enforcement (forensic features), and in machine authentication. In order to decide what features might meet these demands, the committee devised an assessment and evaluation process. In developing this process, the committee concluded that the ideal anticounterfeiting feature has two overarching characteristics: it deters (or prevents) reproduction and it facilitates authentication. These two concepts are at the core of the assessment approach used here. The committee also concluded that a feature differs in deterrence value for each class of counterfeiter, depending on the technology that each employs, as summarized in Table 4-1.

Likewise, a feature's detectability depends on the detection method employed, the nature of the currency transaction, and the education, training, and cognitive capabilities of the recipient of the note. While many transactions take place in a venue in which unassisted human authentication is the preferred (or only) option—such as in a store, restaurant, or taxi—an increasing number of transactions involve alternative assisted methods. These assisted methods might deploy a portable, low-cost, nonproprietary basic device that is generally available and easy to use, or a dedicated machine reader such as an automatic currency acceptor that does not involve a human cash handler. Examples of existing simple devices, of varying effectiveness and popularity, include starch pens, magnifying glasses, and ultraviolet penlights. Given the increasing incidence of human, device-assisted authentication abroad and in the United States, an evaluation of potential new features must include their authentication by simple devices and machines as well as by unassisted humans. Table 4-2 summarizes the leading authentication concepts by type of cash handler—that is, by currency user.

Evaluation Scheme

New feature concepts can span a wide range of technological readiness, from ultralow technology to highly innovative possibilities. Of the many potential feature concepts that emerged from the committee's brainstorming during the course of the study, a subset was ranked for presentation in the report by the feature-assessment process described below. The features evaluated were subsequently

TABLE 4-1 Summary of the Committee's Analysis of the Feature Requirements for the Deterrence of Counterfeiting, by Class of Counterfeiter

Class of Counterfeiter	Technology Used in Counterfeiting	Deterrence Concepts
Opportunist	Commercial home/office digital image technology	Move beyond the reflected image: transmitted light, active features, new materials and devices
Petty criminal	Commercial digital technology, special materials and processes	Require expertise: features that challenge digital reproduction, active features, non-print-based features, novel substrate
Professional criminal	Printing technology, special materials and processes	Raise the stakes: features that require substantial investment to reproduce
State-sponsored	Advanced and capital-intensive technology, forensic features	Use irreversible engineering: proprietary technologies, forensic features that cannot be reverse-engineered

TABLE 4-2 Summary of the Committee's Analysis of Feature Requirements for the Authentication of Currency, by Type of Cash Handler

Cash Handler	Authentication Concepts
General public, unassisted	Ambient examination: visual, tactile, or audible feature that is obvious in various environments. Careful inspection is culturally acceptable.
General public, assisted	The basic device (a "gadget"): obvious signal from a device that is cheap, portable, and easy to obtain—preferably a multiuse item such as a penlight or cellular telephone.
Point-of-sale cashier, unassisted	Low-profile examination: visual, tactile, or audible feature that is obvious in a quick and inconspicuous examination. The appearance of suspicion is not culturally acceptable.
Point-of-sale cashier, assisted	The dedicated device: obvious signal from an inexpensive, single-use device. The acceptable device may vary among retailers, from cheap (starch pen) to moderate (electronic note authenticator).
Bank teller, unassisted	Education and experience: visual, tactile, or audible feature that is obvious to the educated and experienced user. Distinctive "look and feel" is important. Careful inspection is acceptable, but lengthy inspection is time-prohibitive.
Bank teller, assisted	Authenticating device: a dedicated authenticator of moderate to expensive cost. Device must not deface note (that is, no starch pen).
Blind person, unassisted	Beyond the visual: tactile or audible feature with a strong signal.
Blind person, assisted	Portable and reliable device: audible or tactile signal from an inexpensive, portable, dedicated device.
Retail machine reader	Short end first: strong signal from a point sensor in short-end-first feed direction. Handling speed is not critical.
Commercial bank machine reader	Wide end first: strong signal from a point sensor in wide-end-first feed direction, independent of orientation or face. Handling speed is critical.
Central bank machine reader	Beyond authentication: features that provide added information as notes recirculate through the Federal Reserve Banks.
Law enforcement, forensic	Unique and traceable: features that present counterfeiters with many different options regarding simulation or reproduction. These duplicated features can be examined with laboratory equipment to provide a unique "fingerprint" to a particular counterfeiter's work, allowing classification of counterfeiters and traceability of methods and materials.

grouped into intermediate term (less than a 7-year development time anticipated before incorporation into a banknote) and long term (longer than a 7-year development time). The long-term features are described in Chapter 5.

In developing an assessment system, the committee turned to the flow model described in Chapter 2 and recognized that the two most important parameters for assessing the effectiveness of a feature are these: (1) how well the feature deters simulation (decreases ρ) for each class of counterfeiter, and (2) how well it aids in currency authentication (decreases s)—for each class of human cash handler, including the blind, with and without a simple device; for machine readers; and for law enforcement.

Armed with this insight, the committee developed an evaluation matrix that scored the efficacy of each feature for its deterrence and authentication value, by counterfeiter and user, respectively (see Table 4-3). Deterrence value is evaluated for four counterfeiter classes: the opportunist, petty criminal, professional criminal, and state-sponsored counterfeiter. The impact of the primitive counterfeiter is low enough that deterrence is not important for that class.

Validation efficacy is scored for 12 user classes. The general public, cashiers, and tellers are called on to verify currency in ambient conditions, without assistance; thus, unassisted authentication remains important. In addition, these user classes are increasingly employing simple devices to authenticate banknotes in cash transactions. As requested of the committee, the user classes also include the blind (unassisted and assisted by a simple device), machine readers (including automated cashier machines and commercial- and central-bank authenticators), and law enforcement.

When the committee attempted to assign numerical scores to various feature attributes and derive an additive overall score for each feature, it found this scheme inadequate for two important reasons. First, because every selected feature was

TABLE 4-3 Demonstration of the Committee's Evaluation Process for Proposed Innovative Features for Banknotes

	Deterrence to Counterfeiting by:				Validation Efficacy by Type of Currency User											
					Use Unassisted				Use Assisted by Device or Machine							
Innovative Feature Concept	Opportunist	Petty Criminal	Professional Criminal	State-Sponsored	General Public	Cashier (point-of-sale)	Teller (bank-level)	Blind User	General Public	Cashier (point-of-sale)	Teller (bank-level)	Blind User	Retail Automated Cashier	Commercial Bank—Counting, Sorting	Central Bank—Counting, Sorting, Fitness	Law Enforcement—Forensic
Proposed Feature 1	■	■	▩			✔[a]		✔		✔		s	✔	✔	✔	✔
Proposed Feature 2	■	■			✔	✔	✔									

NOTE: "■" indicates features excellent in deterring simulation; "▩," good in deterring simulation; "✔," easy to validate; "s," somewhat easy to validate; and "blank," no value for deterring the counterfeiter class or for authentication by user.

[a]Reflective feature detectable using a collimated light source.

good at something and none was good at everything, the combined score was nearly the same for every feature despite various weighting schemes; consequently, there was little differentiation among features. Second, the flow model outlined in Chapter 2 shows that additive scores are not necessarily relevant to a feature's effectiveness at decreasing the amount of counterfeit currency in circulation. As a result, ranking the features by a numerical score was not a useful exercise. So instead, to score the efficacy values of proposed features, the committee assigned the following symbols: "■" indicates features excellent in deterring simulation, "▩" indicates features good in deterring simulation, "✔" indicates features that would be easy to validate, and "**s**" indicates features that would be somewhat easy to validate. "Blank" in the matrix indicates that the committee saw the feature as having no value for deterring counterfeiting by the class in question or no value for validation by the user in question.

Demonstration of the Evaluation Scheme

Table 4-3 shows an evaluation for two fictional proposed features. Features 1 and 2 are both "excellent" for deterring counterfeiting by opportunist and petty criminal counterfeiters, and Feature 1 is also somewhat effective ("good") for deterring the professional counterfeiter. Table 4-3 shows that Feature 1 is "easy to validate"—that is, very effective for currency-authentication purposes—for seven classes of user. In the case of the general public assisted by a simple device, the device in question would be something like a generic light-emitting diode (LED) penlight or some other sort of simple, filtered, single-wavelength (color) light source. Feature 1 is also "somewhat effective" for blind users—assuming that a device is available to the blind user that would provide an audible or vibratory response to some feature characteristic or the intrinsic tactile nature of the feature would be of use to the blind user unassisted. Feature 2, while effective for fewer users, may be a better choice, as it provides deterrence value to the opportunist and petty criminal counterfeiter while providing unassisted authentication value to the general public, the cashier, and the bank teller—that is, the first line of defense against the passing of counterfeits.

As mentioned above, the committee devised the evaluation scheme to assist in choosing which feature ideas to develop further. For a feature to remain in consideration, it was required to provide excellent deterrence against simulation by at least one type of counterfeiter and to provide excellent authentication assistance by being easy to validate by at least one type of cash handler. Features that did not meet this criterion were eliminated from further consideration. Features were ranked in the matrix exemplified in Table 4-3 on a continuing basis, and as feature ideas evolved, some features dropped off the list and some new features were added. The evaluation scheme, therefore, became a valuable tool for the committee's idea-generation process, as it allows ready identification of features that excel in each category. Furthermore, when used for a variety or combination of features, the evaluation process provides a rationale for selecting features that will offer a broad spectrum of deterrence. While this system does not provide a numerical ranking of any particular feature against another feature, the flow model for counterfeiting discussed in Chapter 2, properly extended and validated, would provide a foundation for a quantitative assessment and comparison of features for a new banknote design.

SUMMARY DESCRIPTION OF NOVEL IMAGE FEATURES

The features summarized in the sections that follow emerged from the committee's generation of ideas and the assessment process described here. A short description of each feature and of the rationale for why it scored high in the evaluation process is supplemented by further details contained in Appendix C, entitled "Intermediate-Term Feature Descriptions." The structure of the appendix is outlined in Box 4-1. The information in Appendix C relating to development risks and issues and to the feature development plan is used as the basis for the discussion in Chapter 6, "A Path Forward."

Looking ahead at this chapter, the results of the committee's evaluation of the deterrence capabilities of the individual features are summarized in Tables 4-4, 4-5, and 4-6. The results of the evaluation on production, discussed in detail in Appendix C for each feature and including a cost assessment, are summarized in Table 4-7. Table 4-8 evaluates the feature concepts against the emerging digital counterfeit technologies that were discussed in Chapter 3. Box 4-2 groups the features by manufacturing technology (printed, substrate, composite additions); Box 4-3 groups the features by primary application (general public unassisted, blind users unassisted, general public assisted with simple device, and machine readers or dedicated device); and Box 4-4 groups features by expected implementation schedule in the near term, within 3 to 4 years, and in the intermediate term, in less than 7 years. The information in Appendix C was used as the basis for the evaluations and descriptions contained in the tables and the boxes.

Overall, these features have considerable potential to deter opportunist and petty criminal counterfeiters. Some features will also have deterrence value for those professional criminal counterfeiters who do not have the resources to adopt the next level of technology. State-sponsored counterfeiters, however, will likely not be deterred by these features because this class of counterfeiter has ready access to technology, including advanced printing and manufacturing processes. Chapter 5 describes disruptive technologies that might be implemented in the longer term—these are currency features that would pose daunting challenges to counterfeiters who rely on conventional manufacturing technology.

BOX 4-1
The Structure of Appendix C, "Intermediate-Term Feature Descriptions"

In-depth descriptions of the innovative banknote features that could be implemented in a time frame of fewer than 7 years are presented in Appendix C of this report and may be referenced for important feature details. Each feature description in Appendix C, "Intermediate-Term Feature Descriptions," includes subheadings dealing with various aspects of the feature:

- *Description*—An explanation of the physical principle(s) on which the feature is based. Also, the feature application as visible, machine-readable, applicable to the visually impaired, forensic applicability, and so on, is described. Furthermore, the benefits and limitations of the feature are presented; graphics may be included to depict the feature and its operation.
- *Feature Motivation*—A summary of the reasons why the feature is highly rated by the committee and reference to its uniqueness.
- *Materials and Manufacturing Technology Options*—A summary of the materials and manufacturing process that could be used to produce the feature, as well as initial thoughts on how the feature could be integrated into a Federal Reserve note.
- *Simulation Strategies*—A discussion of potential ways in which a counterfeiter could simulate or duplicate the feature and the expected degree of difficulty in attempting to do so.
- *Key Development Risks and Issues*—A discussion of the durability challenges, feature aesthetics, anticipated social acceptability, and description of the key technical challenges that must be addressed during the first phase of the development process to demonstrate the feasibility of the feature idea; that is, to demonstrate feature capabilities and determine the usefulness of the feature in counterfeit deterrence. (The development phases are defined in Chapter 6.)
- *Phase I Development Plan*—A characterization of the current maturation level of the feature technology, key milestones to be achieved during the first development phase, and known current and planned related developments external to the Bureau of Engraving and Printing (BEP).
- *Estimate of Production Cost*—An initial assessment of additional BEP operational steps that would be required at the BEP to produce a banknote with the feature, incremental cost (higher, lower, the same) relative to the cost of the current security thread, and an indication of whether additional BEP capital equipment would be required for production.
- *References and Further Reading*—Selected references related to the feature and its associated components. Such references could include, for example, papers and conference proceedings for background on any work done relating to this feature. These lists are not exhaustive but are intended to provide a snapshot of current work related to the feature concept.

The features summarized in this chapter are grouped below in three manufacturing categories:[1] printed features, modified substrate features, and features employing composite additions to the existing substrate. In the first category,

[1]Note that in Appendix C, the features are presented in alphabetical order.

printed features (Table 4-4) are created using conventional printing processes and are intended to be viewed in reflected and/or transmitted light. In the second category, *modified substrate features* (Table 4-5) alter the existing cotton-linen substrate or replace it altogether. The third category, *composite additions to the substrate* (Table 4-6), includes the addition of plastic devices having unique optical and/or tactile properties. These categories were selected because the committee thinks that future anticounterfeiting trends will move away from printed features toward

TABLE 4-4 Committee's Evaluation of Innovative Printed Features for Banknotes

	Deterrence to Counterfeiting by:				Validation Efficacy by Type of Currency User											
					Use Unassisted				Use Assisted by Device or Machine							
Innovative Feature Concept	Opportunist	Petty Criminal	Professional Criminal	State-Sponsored	General Public	Cashier (point-of-sale)	Teller (bank-level)	Blind User	General Public	Cashier (point-of-sale)	Teller (bank-level)	Blind User	Retail Automated Cashier	Commercial Bank—Counting, Sorting	Central Bank—Counting, Sorting, Fitness	Law Enforcement—Forensic
Color image saturation	■	■	▩							✔	✔	✔	✔	✔		✔
Grazing-incidence optical patterns	■	■	▩						✔[a]	s	s					
High-complexity spatial patterns	■	■	▩		✔	✔	s									
Metameric ink patterns	■	■	■						✔[b]	✔	✔	s	✔	✔	✔	✔
Nanocrystal pigments	■	■	■						✔[b]	✔	s					✔
Nanoprint	■	■	■	▩					✔[c]	✔	s					s
See-through registration feature	■	■			✔	s	s									
Thermoresponsive optically variable devices	■	■	■		✔					s	s		s			s

NOTE: "■" indicates features excellent in deterring simulation; "▩," good in deterring simulation; "✔," easy to validate; "s," somewhat easy to validate; and "blank," no value for deterring the counterfeiter class or for authentication by user.

[a]Reflective feature detectable using a collimated light source.

[b]Reflective feature detectable using an LED or filtered light source.

[c]Detectable with clear inscribed plastic sheet to generate moiré pattern of printed image.

TABLE 4-5 Committee's Evaluation of Innovative Modified Substrate Features for Banknotes

	Deterrence to Counterfeiting by:				Validation Efficacy by Type of Currency User											
					Use Unassisted				Use Assisted by Device or Machine							
Innovative Feature Concept	Opportunist	Petty Criminal	Professional Criminal	State-Sponsored	General Public	Cashier (point-of-sale)	Teller (bank-level)	Blind User	General Public	Cashier (point-of-sale)	Teller (bank-level)	Blind User	Retail Automated Cashier	Commercial Bank—Counting, Sorting	Central Bank—Counting, Sorting, Fitness	Law Enforcement—Forensic
Fiber-infused substrate	■	■	■	■	s				✔[a]	✔	✔		✔	✔	✔	s
Microperforated substrate	■	■	▩		s				✔[b]	✔	s					s
Plastic substrate for low-denomination notes	■	■			✔	✔	✔	✔								
Tactile variant substrate	■	■	▩		✔	✔	✔	✔							✔	

NOTE: "■" indicates features excellent in deterring simulation; "▩," good in deterring simulation; "✔," easy to validate; "s," somewhat easy to validate; and "blank," no value for deterring the counterfeiter class or for authentication by user.

[a]Transmitted feature detectable using a device that could range from an LED light source to sophisticated but low-cost instrumentation.

[b]Detectable with transmitted light.

more manipulation of the three-dimensional substrate and the incorporation of composite materials.

The features described in this chapter are expected to be used differently by different cash handlers, including the general public, the blind, cashiers, bank tellers, and machine readers. Some features are intended for unassisted use by cash handlers using direct human perception—that is, the feature is visible in ambient light or is tactilely identifiable. Some features are intended for "assisted use" by cash handlers, meaning that the feature's counterfeit deterrence is enabled by the use of a device—that is, a widely available nonproprietary tool that operates on a feature to produce a human-perceptible signal. Examples of simple devices intended for use by the public include a penlight or a polarizing sheet. Cashiers may use dedicated

TABLE 4-6 Committee's Evaluation of Innovative Features Based on Composite Additions to the Substrate

Innovative Feature Concept	Deterrence to Counterfeiting by:				Validation Efficacy by Type of Currency User: Use Unassisted				Use Assisted by Device or Machine							
	Opportunist	Petty Criminal	Professional Criminal	State-Sponsored	General Public	Cashier (point-of-sale)	Teller (bank-level)	Blind User	General Public	Cashier (point-of-sale)	Teller (bank-level)	Blind User	Retail Automated Cashier	Commercial Bank—Counting, Sorting	Central Bank—Counting, Sorting, Fitness	Law Enforcement—Forensic
Fresnel lens for microprinting self-authentication	■	■			✔	s	s	s								
Hybrid diffractive optically variable devices	■	■	■		✔	s	s	s		✔	✔	s	✔			
Refractive microoptic arrays	■	■	■		✔	s	s	s								s
Subwavelength optical devices	■	■	■	▨	s				✔[a]	✔[a]						s
Window	■	■	■		✔	s	s	✔							✔	

NOTE: "■" indicates features excellent in deterring simulation; "▨," good in deterring simulation; "✔," easy to validate; "s," somewhat easy to validate; and "blank," no value for deterring the counterfeiter class or for authentication by user.

[a]Detectable with a thin-film polarizer or polarized light source such as illumination from cellular telephone display.

devices designed and sold for the sole purpose of banknote authentication but that are still inexpensive and easy to use; examples include electronic devices that read a portion of the banknote and indicate its authenticity and denomination. These electronic devices would also be of use by the blind. Features that are effective for automated retail purposes and for commercial banks would be expressly designed for machine readers. The committee made a preliminary assessment of which features would be readily detected by devices and sophisticated machine readers; as various features are further explored during a research and development program, a key task could further explore the best methods of assisted detection. Finally, features effective for forensics would provide law enforcement with additional capability for associating particular counterfeiting methods with certain counterfeiters.

Printed Features

Color Image Saturation

The color image saturation feature embeds information in the FRN by varying the intensity, or "saturation," of colors in a printed image in a prescribed way. The printed variations in color intensity can be subtle, so as to prevent notice by the unaided eye. Digital scanners are unable to capture the variations of the feature because they quantize the recorded intensity. However, these genuine variations in color intensity can be readily detected by a dedicated optoelectronic device that can determine whether or not a banknote is authentic. Digital scanners may be used for the analysis of this feature. However, most digital scanners record red, green, and blue channels of information, and luminance is not a signal channel in most scanners, although the red, green, and blue data can be processed to produce a luminance signal using known colorimetric methods. Depending on the subtlety of the embedded luminance signal, 8-bit signals may or may not be sufficient.

Grazing-Incidence Optical Patterns

The grazing-incidence optical pattern feature uses deformations in the surface of a note to generate diffraction patterns that are clearly visible when illuminated and/or viewed at high angles of incidence; that is, the illuminating light is incident on the currency surface at angles greater than 80 degrees. The deformations required for this feature could be added by the intaglio process, exploiting the capability of the high-pressure intaglio presses. This process cannot be easily duplicated by the digital printing processes discussed in Chapter 3. This feature could be observed directly, or with the aid of a simple device such as a penlight.

High-Complexity Spatial Patterns

The high-complexity spatial patterns feature relies on a highly dense and specially designed printed pattern that exploits weaknesses in electronic printer capabilities that are not shared with the conventional analog printing methods currently used for currency production. Thus, for properly designed patterns, the commonly available electronic printers such as ink-jet and other printers will not be able to reproduce defect-free images. The counterfeit images would be readily apparent to the unaided eye.

A representative high-complexity spatial pattern is depicted in Figure 4-1. In this figure, the center of the pattern is indistinct because of the printing process used to create this report. On an authentic banknote, the radial lines would continue to the extreme center of the pattern, providing a much crisper image. The

FIGURE 4-1 A starburst pattern is an example of the kind of high-complexity spatial pattern that electronic printers could not reproduce.

committee is not suggesting that this particular pattern be used. However, the gradually decreasing spacing, whether radial or lineal, in patterned images such as this will at some point cause the addressability limits of an electronic printer to place two lines together without a space, or two spaces together without a line, producing a clearly inferior image. With proper design and adequate intaglio printing quality, the analog system should be consistently able to frustrate a digital system's ability to duplicate the image.

Metameric Ink Patterns

Metamerism results when two colors appear to be the same when they are viewed using one wavelength (or color) of light, but appear to be different when viewed under another wavelength. In commercial printing, in which special inks are often used, care is taken to avoid metameric effects. In this instance, printer specialists become frustrated when two nearby colors appear to be the same color under one light source—say a fluorescent lamp—and then appear as different colors under a conventional tungsten lamp. However, this printing "problem" can be exploited as a feature, possibly coupled as a further enhancement with nanocrystal inks, to generate patterns that can be observed under the special illumination conditions of two different light colors or with a simple filter. Furthermore, a sophisticated authentication test using a dedicated optoelectronic device can be easily developed to measure the reflectance spectrum or "fingerprint" of the metameric inks. Special filters coupled with controlled illuminants may also be used for readily viewing the patterns in the metameric inks. By using special inks for the printing of features on banknotes, such as those enabled by nanocrystal technology (see the discussion below), counterfeiters would not be able to produce the same effect with good quality when using other "normal" inks. Miniature instrumentation could be readily built that would clearly discriminate between the "real" nanocrystal inks and the simulated attempts.

Nanocrystal Pigments

Dispersions of semiconductor or metallic nanocrystals have the potential to provide unique color and spectral characteristics in an ink format that could be used to form images on currency or, as embedded in a plastic matrix, to create colored features in a security strip. By synthesizing nanocrystals for specific ink formulations, complex color characteristics can be achieved with a degree of control and durability that would be difficult to duplicate using other approaches. These inks could also be used for novel metameric applications, as described above. This feature would require a dedicated optoelectronic device for measuring the color spectrum of the printed feature.

Nanoprint

Microprinting, currently printed on the FRN's paper substrate, can be viewed with a very sharp eye or modest optical aids. Nanoprint involves printed text, images, or regular arrays of patterns with critical dimensions in the submicron range, formed on smooth surfaces of components of the currency, such as the security strip. Lithography research has produced several methods that can "print" structures and patterns with dimensions in the deep submicron and nanometer regime. Nanoprint rated highly in the committee's evaluation process because it provides the combination of an extremely high level of security with the potential for low cost. Nanoprint can be viewed directly using a very high magnification electron microscope, but these are only available in specialized laboratories. For use by cash handlers, nanoprinted features would be designed to generate diffraction, moiré, or other patterns. These features would be observed indirectly by exploiting visible and/or machine-readable collective effects such as diffractive or moiré effects.

See-Through Registration

The see-through registration feature involves placing on one side of the currency substrate a printed image that works in tandem with an additional element printed on the opposite side of the currency. When the two images are well aligned and the note is held up to the light, the complete (fully aligned) feature is visible through the substrate. The two images on either side of the note can also combine to create a transmitted color different than either of the printed images' colors. This feature is robust and is easily visible. It is currently used in several types of currency, and appears to be effective. It can be implemented in the near term. In the longer term, advances in reprographic technology may make this feature easier to duplicate.

Thermoresponsive Optically Variable Devices

Thermoresponsive optically variable devices use printed inks or other materials that change in appearance with temperature. For example, the inks used to print such a feature might appear blue when cool and red when hot. Figure 4-2 illustrates the response of such a thermally sensitive material to heat from the human hand. This effect is reversible, meaning that after becoming hot and hence red, the color of the feature can return to being blue when cool. A counterfeiter would have to obtain and color-match the appropriate liquid-crystal ink to replicate this feature.

Modified Substrate Features

Fiber-Infused Substrate

The fiber-infused substrate feature embeds small-diameter optical-fiber segments, such as optical plastic, in the currency substrate. These fibers scatter light in a unique way, providing a kind of fingerprint for the currency substrate that is difficult to reproduce. The effect could be detected in transmitted light or induced

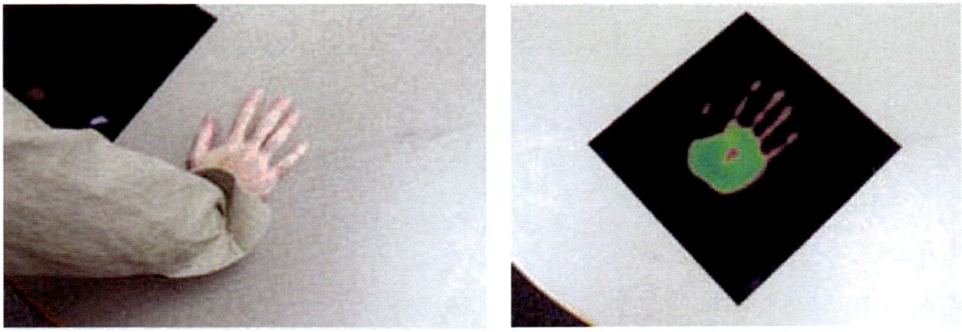

FIGURE 4-2 Images showing the thermal response of a cholesteric liquid crystal. At room temperature, the device is black (upper left in the left frame). Heating by a hand causes changes in the observed color (right frame). SOURCE: Adapted from the University of Wisconsin-Madison IPSE Liquid Crystal Activity Guide; <http://mrsec.wisc.edu/Edetc/supplies/ActivityGuides/LC_Activity_Guide_Expo.pdf>.

using a small penlight. This feature would not be reproducible using electronic printing and scanning techniques. It can be considered an upgrade to the red and blue fibers currently used in U.S. banknotes. When viewed by an instrument that illuminates the fibers and collects the returned optical signal, the unique fiber-absorption characteristics and other optical signatures would be extremely difficult to replicate and hence would provide for excellent authentication.

Microperforated Substrate

The microperforated substrate feature employs very small holes punched into the currency substrate to form symbols or denomination indications that would be easily detected in transmitted light. A simple penlight could be used to make this feature more apparent. Specialized perforating machinery is required to make these small holes without tearing or otherwise damaging the currency substrate. Depending on the cost and performance requirements, the perforations might also be generated by short-pulse laser systems such as femtosecond lasers, which can produce very small holes in many materials ranging from steel to plastic, glass, and even diamonds. Furthermore, the high pulse rate of such laser systems could permit fast generation of the notes.

Plastic Substrate

Although plastic has been historically examined as a matrix for banknote production, it has only become a viable alternative medium during the past decade. In 1996, Australia became the first country to convert fully to polymer-based banknotes, after having experimented with synthetic polymers notes since 1988. One of the more frequently employed synthetic polymers used for these types of banknotes is biaxially oriented polypropylene.

Proponents of these relatively new banknote materials have pointed out performance benefits that include cleanliness, durability, processability, recyclability, security, and compatibility with existing printing equipment and inks. Other experts have indicated that plastic banknotes are more costly to produce. As of 2006, plastic substrates are used by a number of countries, including Bangladesh, Brazil, Brunei, Chile, China, Indonesia, Kuwait, Malaysia, Mexico, Nepal, New Zealand, Northern Ireland, Papua New Guinea, Romania, Singapore, Solomon Islands, Sri Lanka, Taiwan, Thailand, Vietnam, Western Samoa, and Zambia.

The largest drawback with the use of an alternate substrate in U.S. banknotes is that the plastic substrate would have a tactile feel much different than the traditional Federal Reserve note. If the committee heard one common theme from counterfeit security experts, it was that the distinctive tactile feel of U.S. banknotes is the foremost feature used by most people to judge the authenticity of U.S. currency. This is true according to anecdotal stories (for example, the first person to detect a $100 supernote was a cash handler who thought the bill "didn't feel right"), focus-group studies, and controlled scientific studies.[2] The committee believes that the best security features are those that are easiest to detect. Therefore, there must be an indisputably compelling reason to make a change in the tactile feel of the substrate that arguably is the easiest and most recognized FRN feature. Also, a significant public education program would be necessary to introduce such a change, and during the interim period before the alternate substrate became widely known and accepted, it is feared that counterfeiting would flourish owing to confusion about the genuine substrate. Unlike many other countries, the United States never recalls old banknotes, so confusion regarding the substrate could persist for some time. Given that the committee was unable to construct a compelling case for the counterfeit-deterrence value of a wholesale change to an alternate substrate, the plastic substrate is not included on the list of candidate technologies.

However, the committee considered another possibility that would involve a partial change. The plastic substrate could be used for low-denomination notes. This would provide a significantly different tactile feel for use in denominating by the blind, and it would complicate the practice of bleaching low-denomination notes to produce a quality substrate for higher-denomination counterfeits. The improved durability of the plastic substrate would also provide a cost benefit, since low-denomination banknotes suffer the most user wear and have shorter lifetimes in the marketplace owing to higher usage rates in cash transactions.

It has been reported that the lifetime of plastic banknotes can be as great as four times that of paper.[3] The life span of U.S. currency was examined in a comprehensive 2002 Federal Reserve Board (FRB) study that addressed the applicability of alternative substrates.[4] The report highlighted the fact that the life of the $1 note was more than 20 months in 2001, which is a longer duration than the life span of comparable denominations in several other countries for which the data are available. These data indicate that a plastic substrate would provide about a twofold lifetime increase for U.S. currency. The conversion of the paper $1 notes to a plastic substrate would require a large investment to accommodate, estimated

[2]H. Treinen. 2004. Banknote recognition: Public reception of double sided intaglio printed banknotes compared with single sided intaglio and offset counterfeited banknotes. Currency Conference, Rome, 2004. Personal communication.

[3]For additional information, see the paper by L. Coventry, 2001, Polymer notes and the meaning of life, presented at 2001 Reserve Bank of Australia, Currency Conference, Barcelona, Spain, April 2001. Available at <http://www.rba.gov.au/CurrencyNotes/ConferencePapers/cu_coventry_0401.pdf>. Accessed February 2007.

[4]Michael Lambert, Marsha Reidhill, Genie Foster, and Doug Rodriguez. 2002. Lifespan of U.S. currency and analysis of an alternate substrate, unpublished paper presented at the 2002 Currency Conference and 2002 CEMLA Conference.

by the authors of the FRB report to be on the order of $100 million. The results of a Monte Carlo simulation of plastic substituted for the paper substrate indicated that the lifetime of the plastic substrate would have to be greater than 36 months for the replacement to be cost-effective. The available lifetime data for a plastic substrate is approximately at this level.

Future reassessment of the cost-effectiveness of a plastic substrate could be done after additional lifetime data become available. Any future study that evaluates the replacement of low-denomination paper-based banknotes (that is, the $1, $2, and $5 FRN) with plastic should also assess the counterfeiting-protection benefit acquired from converting to plastic, including the public confusion in handling two substrates that "feel" different.

If the conversion to a plastic substrate is eventually made for low-denomination notes, that experience will provide useful information for the further assessment of converting all FRNs to the alternate substrate at some later time.

In this chapter, the committee identifies some features that would be added to the current substrate to produce a hybrid substrate, such as the Fresnel lens and window. And Chapter 5 discusses the anomalous currency space that would serve as a localized region within the current paper substrate for inserting plastic, substrate-based features that would be selected on the basis of providing maximum counterfeiting deterrence. This approach would locally change the tactile feel of the banknote.

Tactile Variant Substrate

Tactile variance features create changes in the local "feel" of a banknote across the surface of the currency. This variance can be introduced in a number of ways, such as by incorporating a variety of security threads with different embossing and/or at different locations on different denominations. This feature would deter the bleaching of FRNs in order to use the substrate to make other banknotes of higher denominations. Providing a tactile feel that varies by denomination provides a means for the visually impaired to denominate U.S. banknotes.

Composite Additions to the Substrate

Fresnel Lens for Microprinting Self-Authentication

The Fresnel lens for microprinting self-authentication feature provides an on-the-banknote optical aid for reading the fine features printed on the substrate. The lens would be an authenticating feature used by bending the note to inspect fine detail of microprinted features through it. The feature consists of a thin plastic lens that is incorporated into the substrate. The plastic lens could also impart a distinct tactile feel to the note and thus serve as a tactile denominating feature for a blind person if the lens placement were denomination-specific. This feature would be difficult to simulate with high quality because of challenges presented by fabricating a sufficiently thin device and integrating it into the substrate.

Hybrid Diffractive Optically Variable Devices

Conventional diffractive optically variable devices, such as holograms, are in widespread use, as are nondiffractive components, such as optically variable inks. The proposed hybrid diffractive optically variable device combines diffractive and nondiffractive effects into one single device. For instance, polarization-dependent effects and encrypted images can be combined with holograms and interference structures to provide a visibly striking image in a single feature. Also, the optical phase and other information in the images produced by the devices can be encoded and retrieved using specialized equipment that is very difficult for even sophisticated replicas to replicate.

Refractive Microoptic Arrays

The refractive microoptic array feature class employs microoptic element arrays of tiny lenses that are integrated into the substrate to create obvious interactive visual effects. As the user moves or tilts the currency, the image being observed changes. The visual effect is striking and easily seen. This feature has a tactile component that might help the blind authenticate an FRN without device assistance. Even professional counterfeiters would have significant difficulty reproducing the optical effects because of challenges presented by fabricating a sufficiently thin device and integrating it into the substrate. For example, a hot-stamped holographic strip might yield an optically variable feature, but it would lack the physical integration of the optical material with the substrate.

Subwavelength Optical Elements

A wide range of optical characteristics can be obtained from a single material by structuring it on length scales shorter than the wavelength of light—that is, substantially less than a few hundred nanometers. For example, parallel arrays of thick, subwavelength metal lines on a transparent substrate are completely transparent to light polarized along the length but are highly opaque to light with the orthogonal polarization. Figure 4-3 shows one example. Features based on subwavelength optical elements could provide visible images that produce distinctive colors, polarization contrast, and other optical effects. They can be used to provide a high level of control over the appearance of an image. While some of the effects

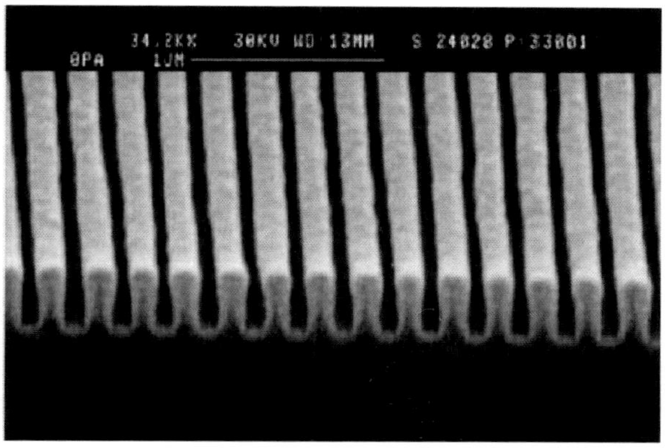

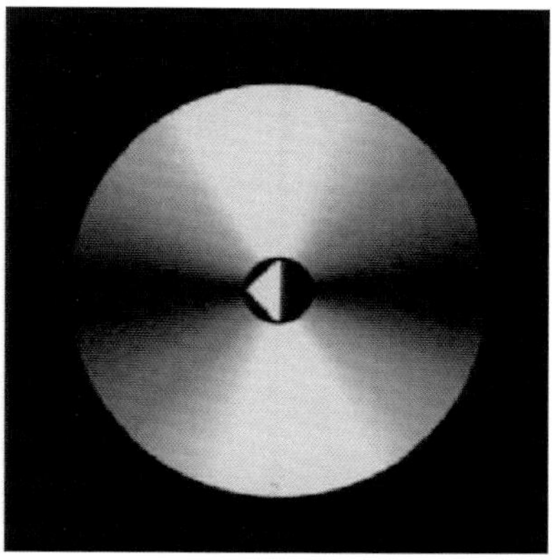

FIGURE 4-3 Scanning electron micrograph (left) of a subwavelength optical device, which in this case consists of an array of metal lines. This structure forms a polarizer, whose implementation as a circular polarizing element is illustrated on the right. SOURCE: Reprinted, by permission, from B. Schnabel, E.-B. Kley, and F. Wyrowski, 1999, Study on polarizing visible light by subwavelength-period metal-stripe gratings, Optical Engineering 38(2): 220-226.

would be visual in unaided examination, the effects would be more dramatic when examined using a polarizing film.

Window

Windows can be integrated into the substrate using a different shape and location for each denomination. For example, if the windows were progressively smaller with each denomination, they would complicate the bleaching of low-denomination notes for use in high-denomination counterfeits. Different denominations could also have different window shapes and positions. Their tactile properties also could be used by the blind for determining denominations. Note that substrate integration technology similar to that required to implement windows is also needed for the Fresnel lens, the hybrid diffractive, the microoptic array, and the subwavelength optical element features. It is envisioned that in the future a window could also be used as a platform for a wide variety of sophisticated features, as described in the section entitled "Anomalous Currency Space" in Chapter 5 and Appendix D.

ASSESSMENT OF FEATURE PRODUCTION COST

The committee made a preliminary estimate of each innovative feature's manufacturing cost on the basis of a preliminary understanding of the feature technology and limited knowledge of the Bureau of Engraving and Printing's (BEP's) production capability. Table 4-7 summarizes the committee's production cost assessments, described in Appendix C.

The committee's cost assessment does not explicitly take into account the new equipment that feature suppliers would have to acquire, nor that of the substrate supplier for those features that would be integrated within the substrate. It is assumed that these costs would be built into the feature cost and thus are captured as the incremental cost of producing the new feature, compared with the cost of

TABLE 4-7 Production Cost Assessments of Innovative Features for Banknotes

Innovative Feature Concept[a]	Requires Additional Bureau of Engraving and Printing (BEP) Operational Steps[b]	Incremental Cost Relative to the Security Thread[c]	Requires New Capital Equipment at BEP[d]
Color image saturation		O	
Fiber-infused substrate		+	?
Frensel lens for microprinting self-authentication		+	
Grazing-incidence optical patterns	?	O	?
High-complexity spatial patterns		O	
Hybrid diffractive optically variable devices	?	+	?
Metameric ink patterns		O	
Microperforated substrate	●	O	●
Nanocrystal pigments		O	
Nanoprint		+	
Refractive microoptic arrays		+	?
See-through registration		O	
Subwavelength optical elements		+	
Tactile variant substrate		O	
Thermoresponsive optically variable devices	?	+	●
Window	?	O	?

[a]Can be implemented in less than 7 years.
[b]"●," step required; "?," possible new step required, depending on implementation.
[c]"O," roughly equivalent to current costs; "+," more costly.
[d]"●," new equipment required; "?," possible new equipment required, depending on implementation.

producing the current security thread. The cost of developing devices for detecting machine-readable features also is not included in these estimates.

On the basis of this preliminary assessment, all of the intermediate-term features that the committee has described appear to be viable candidates for further consideration, with the possible exception of the plastic substrate for the $1 FRN owing to the Federal Reserve's estimate of $100 million required for infrastructure changes.[5] As discussed earlier in this chapter, the cost of implementing a plastic substrate may not be adequately offset by the added durability advantage and counterfeit-deterrence benefit provided by eliminating a counterfeiter's option of bleaching and printing a larger-denomination banknote on the lower-denomination substrate.

PRIORITIZATION OF INNOVATIVE FEATURES

The features described in this chapter incorporate a wide range of potentially effective technologies. The committee's intention is to suggest a number of feature ideas that, once developed, will provide the banknote designer with a portfolio of innovative counterfeiter-deterrent features. Since not every feature idea will be developed, the prioritization of features for entry into a development program would take many factors into account. Those factors would include, for example, the estimated deterrent effectiveness against the projected information technology capability, a balanced mix of printed image and substrate features, a high degree of recognizability by the different types of cash handlers using sensory perception or assisted by some type of equipment, and the expected development time line. These considerations, matched against the BEP's internal assessment of critical security-feature needs and metrics, would highlight those features for top development priority. The features can be combined in different ways to provide a layered defense against the various classes of counterfeiter. Ultimately, then, the prioritization of the features would be accomplished by the banknote designers.

Effectiveness against evolving counterfeiting threats is a prime consideration when prioritizing potential features for development. Table 4-8 summarizes the committee's evaluation of the deterrence value of the innovative security features with respect to various digital image technologies that could be used in counterfeiting. This analysis is based on the committee's estimation of the difficulty in applying these advanced digital copying and reprographic technologies to simulate the proposed features. A comparison with Table 3-1, which evaluates the deterrence value of current FRN features against the current digital threat, indicates a

[5]Michael Lambert, Marsha Reidhill, Genie Foster, and Doug Rodriguez. 2002. Lifespan of U.S. currency and analysis of an alternate substrate, unpublished paper presented at the 2002 Currency Conference and 2002 CEMLA Conference.

TABLE 4-8 Committee's Evaluation of Innovative Security Features in Deterring Digital Age Counterfeiting

Features	Counterfeit-Deterrence Characteristic	Advanced Digital Printer	Advanced Color Copier	Scanners and Digital Cameras	Imaging Software	Potential for Attack Through Internet Information Exchange
Printed Features						
Color image saturation	Digital signature embedded in currency	●	●	●	●	✖
Grazing-incidence optical patterns	Exploitation of three-dimensional nature of intaglio printing to emboss an optical feature	●	●			✖
High-complexity spatial patterns	Challenge to all digitally addressable printing	●	●	●	●	
Metameric ink patterns	Specialized printing dye mixtures yield indistinguishable visual color, yet easily differentiated with simple instrumentation	●	●	●	●	✖
Nanocrystal pigments	Nanocrystals and dyes with brightness and unique colors and fluorescence	●	●	●	●	✖
Nanoprint	Print smaller than any commercial printing methods	●	●	●	●	
See-through registration	Taking advantage of dual-side printing at BEP	●	●	●		✖
Thermoresponsive optically variable devices	Colors that change with temperature	●	●	●		✖
Modified Substrate Features						
Fiber-infused substrate	Transmitted effect	●	●	●		
Microperforated substrate	Transmitted effect	●	●	●	●	
Tactile variant substrate	Patterned change in the feel of the surface	●	●	●	●	

Plastic substrate for low denominations	Change in texture and mechanical properties	●	●	●		✖
Composite Substrate Features						
Fresnel lens for microprinting self-authentication	Optical device that is built into substrate to assist in authentication of microprint	●	●	●	●	
Hybrid diffractive optically variable devices	Combining of holographic features with other advanced optical features	●	●	●	●	✖
Refractive microoptic arrays	Optical device for built-in authentication	●	●	●	●	
Subwavelength optical elements	Very-high-resolution printing that produces a patterned optical device or effect that is unable to be made in another way	●	●	●	●	✖
Window	Denomination-specific window with distinct tactile feel	●	●	●		

NOTE: Entries in the "Counterfeit-Deterrence Characteristic" column recall the strength of the feature in question. "●" indicates that the feature has deterrent value for the identified image technology. The "✖" in the column "Potential for Attack Through Internet Information Exchange" indicates a vulnerability of the feature to the rapid information exchange, enabled by the Internet, of data, images, and methods of simulation.

marked improvement in effectiveness for all of the innovative features, with the caveat that these new features must be further developed to fulfill the expectation. The resistance of the features to duplication would be measured during the development program. Table 4-8 also indicates which features may be vulnerable to information exchange via the Internet—in this case, the threat would consist of information sharing of details helpful to counterfeiters, such as a "how to" guide on simulating a feature, specific sources through which critical materials are available, a graphical representation of a feature or portion of a banknote that can be downloaded, and so on.

Box 4-2 groups the features by manufacturing technology as printed, substrate, and substrate composite additions features. Five features incorporate the use of additional materials onto the banknote to produce a composite structure. The committee forecasts that the improvements in digital printing technology will require less reliance on printed images for counterfeit deterrence and more reliance on composite features embedded in the substrate. These composite features would

BOX 4-2
Groupings of Innovative Banknote Security Features by Manufacturing Technology

- Printed Features
  —Color Image Saturation
  —Grazing-Incidence Optical Patterns
  —High-Complexity Spatial Patterns
  —Metameric Ink Patterns
  —Nanocrystal Pigments
  —Nanoprint
  —See-Through Registration Feature
  —Thermoresponsive Optically Variable Devices

- Substrate Features
  —Fiber-Infused Substrate
  —Microperforated Substrate
  —Plastic Substrate for Low-Denomination Notes
  —Tactile Variant Substrate

- Substrate Composite Additions
  —Fresnel Lens for Microprinting Self-Authentication
  —Hybrid Diffractive Optically Variable Devices
  —Refractive Microoptic Arrays
  —Subwavelength Optical Devices
  —Window

employ special materials and advanced manufacturing processes in order to deter counterfeiters effectively.

Box 4-3 groups the features by application, according to the class of cash handler that the feature is targeting. Readily identifiable features relying entirely on sensory perception are highly desired and were a significant focus of the commit-

BOX 4-3
Groupings of Innovative Banknote Security Features by Application

- General Public Unassisted
  —Fresnel Lens for Microprinting Self-Authentication
  —High-Complexity Spatial Patterns
  —Hybrid Diffractive Optically Variable Devices
  —Plastic Substrate for Low-Denomination Notes
  —Refractive Microoptic Arrays
  —See-Through Registration Feature
  —Tactile Variant Substrate
  —Thermoresponsive Optically Variable Devices
  —Window

- General Public Assisted with Simple Device
  —Fiber-Infused Substrate
  —Grazing-Incidence Optical Patterns
  —Metameric Ink Patterns
  —Microperforated Substrate
  —Nanocrystal Pigments
  —Nanoprint
  —Subwavelength Optical Devices

- Blind Users Unassisted
  —Fresnel Lens for Microprinting Self-Authentication
  —Hybrid Diffractive Optically Variable Devices
  —Plastic Substrate for Low-Denomination Notes
  —Refractive Microoptic Arrays
  —Tactile Variant Substrate
  —Window

- Machine Reader or Dedicated Device
  —Color Image Saturation
  —Fiber-Infused Substrate
  —Hybrid Diffractive Optically Variable Devices
  —Metameric Ink Patterns

NOTE: Some features have multiple applications.

tee. However, as counterfeiting technology continues to improve, the committee foresees an increased interest by cash handlers in using simple devices to aid in the authentication of banknotes. This trend could also capture the interest of the public, so that simple, readily accessible devices could be commonly used to validate a banknote. The grouping includes the following, among other features:

- *General public, unassisted:* Fresnel lens for microprinting self-authentication embedded in the paper substrate, high-complexity spatial patterns, refractive microoptic arrays, microperforated substrate, see-through registration feature, tactile variant substrate, and thermoresponsive optically variable devices.
- *General public, assisted with simple device:* Fiber-infused substrate, grazing-incidence optical patterns, metameric ink patterns, nanocrystal pigments, nanoprint, and subwavelength optical elements.
- *Blind users unassisted* (features that provide tactile information to aid in denomination): Fresnel lens for microprinting self-authentication embedded in the paper substrate, refractive microoptic arrays, tactile variant substrate, and window.
- *Machine readers or dedicated devices:* Metameric ink patterns and color-image saturation.

Box 4-4 lists those features that the committee estimates can be ready for implementation in the near term, within 3 to 4 years, and those that could require an intermediate term, up to 7 years of development. Of the features described in this chapter that are image-based printed features employing conventional inks, the committee considers that three could likely be implementable in the near term:

- Grazing-incidence optical patterns,
- High-complexity spatial patterns, and
- A see-through registration feature.

Four features are printed using specialty inks; one of those could be implemented in the near term:

- Metameric ink patterns.

Three features modify the substrate itself; one of those could be fully developed in the near term:

- The microperforated substrate.

BOX 4-4
Groupings of Innovative Banknote Security Features by Expected Implementation Schedule

- Near Term (within 3 to 4 years)
  —Grazing-Incidence Optical Patterns
  —High-Complexity Spatial Patterns
  —Metameric Ink Patterns
  —Microperforated Substrate
  —Plastic Substrate for Low-Denomination Notes
  —See-Through Registration Feature

- Intermediate Term (within 7 years)
  —Color Image Saturation
  —Fiber-Infused Substrate
  —Fresnel Lens for Microprinting Self-Authentication
  —Hybrid Diffractive Optically Variable Devices
  —Nanocrystal Pigments
  —Nanoprint
  —Refractive Microoptic Arrays
  —Subwavelength Optical Devices
  —Tactile Variant Substrate
  —Thermoresponsive Optically Variable Devices
  —Window

With the exception of the all-plastic substrate, it is assumed that for the features described in this chapter the substrate will continue to be the current cotton-linen blend for U.S. banknotes. The committee believes that the legacy substrate contributes substantially to current authentication—"feel" is the foremost feature used by most people to judge the authenticity of U.S. currency, and controlled access to this substrate is an effective deterrent to opportunist counterfeiters.[6] Although the current paper substrate is distinctive, durable, and effective for authentication, chemical bleaching and/or the de-inking of lower-denomination genuine notes provides the counterfeiter an excellent substrate with which to manufacture higher-denomination notes. Several new features could directly address this threat: those with a composite addition to the substrate, such as a window, and low-denomination plastic notes. The key to the value of the composite additions is that

[6]L. DiNunzio and S.E. Church. 2002. Evaluating public awareness of new currency design features. Proceedings of SPIE, Optical Security and Counterfeit Deterrence Techniques V, R.L. van Renesse (ed.), Vol. 4677, pp. 1-14.

the feature must be progressively different on increasing denominations. Similarly, low-denomination notes made from plastic could not be recycled as $20 or $50 notes, since the plastic substrate has a markedly different look and feel from the cotton-linen substrate. However, changing the substrate material may not provide an overall increase in counterfeit deterrence, as discussed earlier.

CONCLUSIONS

- A feature's deterrence value varies with the class of counterfeiter. A feature's authentication requirements differ with the type of cash handler.
- An additive numerical model for feature evaluation is not optimal, as it lacks the ability to distinguish feature strengths adequately and cannot be extended to estimate the resulting reduction of counterfeit currency in circulation.
- A nonlinear evaluation method—such as one based on a model that incorporates consideration of excellence in (1) deterring particular classes of counterfeiters and (2) aiding authentication for specific types of users—can provide a useful tool for prioritizing features.
- The optimal set of features included on a banknote would address all counterfeiter and user classes.
- Seventeen features for U.S. banknote applications were identified by the committee as deterrents to anticipated counterfeiting threats. Five of these features (metameric ink patterns, see-through registration features, microperforated substrates, plastic substrates, and windows) have been employed already in currency applications in other countries and thus could be implemented more quickly than the other novel features. Each new feature has the potential to serve as an excellent deterrent for at least one class of counterfeiter, although none is expected to deter state-sponsored counterfeiters completely.
- Some of the proposed features are readily human-perceptible, while others require the use of readily available devices. In the future, the committee foresees that the use of devices will be common.
- In general, the committee concludes that the trend is away from reliance on the two-dimensional image, a trend that the Bureau of Engraving and Printing has already begun with the use of a security thread, watermarks, and optically variable ink.
- The committee suggests the need for an increased use of composite additions to the substrate to thwart the threats posed by advanced reproduction devices.

5

Disruptive Feature Platforms

In developing ideas for possible technologies for anticounterfeiting features, there will always be technologies that can be applied in the near term as well as those that require a longer-term approach. This chapter addresses what the committee refers to as feature platforms, which require a longer time scale, more than 7 years, for implementation in currency. These feature platforms will require additional evaluation, research, and development in order to mature salient portions of the technology so as to allow the identification of specific counterfeit-deterrent features that could be effectively implemented into banknotes. Specific details of these feature platforms are described in detail in Appendix D, "Long-Term Feature and Feature Platform Descriptions." Box 5-1 outlines the information presented in this appendix for each feature platform.

DEVELOPING THE IDEA OF FEATURE PLATFORMS

Currency is now produced almost exclusively by conceptually old printing techniques (for example, offset, intaglio, and so on) that have traditionally dominated the conventional printed-paper industry. Also, most if not virtually all of the banknote anticounterfeiting features have been optical. The nature of emerging threats and the even more capable digital technologies that are becoming available to the counterfeiter suggest that this basic approach to counterfeit deterrence will likely not continue to provide a path to secure currency indefinitely into the future. Intrinsic limits in resolution, registration, pattern layouts, and inks associated with these printing methods may make it necessary to consider radically different materials and manufacturing concepts.

BOX 5-1
The Structure of Appendix D, "Long-Term Feature and Feature Platform Descriptions"

The in-depth feature descriptions of the longer-term features and feature platforms—that is, features that can be implemented in a time frame of more than 7 years—are presented in Appendix D, "Long-Term Feature and Feature Platform Descriptions."

- *Description*—An explanation of the physical principle(s) on which the feature is based. Also, the feature application as visible, machine-readable, applicable to the visually impaired, forensic applicability, and so on is described. Furthermore, the benefits and limitations of the feature are presented; graphics may be included to depict the feature and its operation.
- *Feature Motivation*—A summary of the reasons why the feature is highly rated by the committee and reference to its uniqueness.
- *Potential Implementations*—A description of scenarios that provide examples of how the feature could be employed to deter counterfeiting.
- *Materials and Manufacturing Technology Options*—A summary of the materials and manufacturing process that could be used to produce the feature as well as initial thoughts on how the feature could be integrated into a Federal Reserve note.
- *Simulation Strategies*—A discussion of potential ways in which a counterfeiter could simulate or duplicate the feature, and the expected degree of difficulty in attempting to do so.
- *Key Development Risks and Issues: Phase I*—A discussion of the durability challenges, feature aesthetics, anticipated social acceptability, and description of the key technical challenges that must be addressed during the first phase of the development process to demonstrate the feasibility of the feature idea: that is, demonstrate feature capabilities and determine the usefulness in counterfeit deterrence. (The development phases are defined in Chapter 6.)
- *Development Plan: Phase I*—A characterization of the current maturation level of the feature technology, key milestones to be achieved during Phase I, known current and planned related developments external to the Bureau of Engraving and Printing (BEP), and a high-level schedule for Phase I.
- *Estimate of Implemented Production Costs*—An initial assessment of additional BEP operational steps that would be required at the BEP to produce a banknote with the feature, incremental cost (higher, lower, the same) relative to the cost of the current security thread, and an indication of whether additional BEP capital equipment would be required for production.
- *References and Further Reading*—Selected references relating to the feature and its associated components. Such references could include, for example, papers and conference proceedings for background on any work done relating to this feature. These lists are not exhaustive but are intended to provide a snapshot of current work related to the feature concept.

The revolutionary, longer-term approaches suggested in this chapter exploit a wide range of technologies. Certainly, features based on the optical properties of materials will remain highly utilized since currency is for the most part a visual medium. But the committee concluded during the study's feature brainstorming and evaluation process that advances in currency features will progress from a reliance on printed optical features to a reliance on features based on advanced materials and processes that would be far above the simulation capability of opportunist counterfeiters. Indeed, many of these advanced features could only be plausibly attempted by state-sponsored counterfeiters with great effort.

The feature platforms presented in this chapter require the use of special materials and/or special manufacturing technologies, such as microfabrication. These new production platforms can produce highly recognizable features that would be very challenging for a counterfeiter to replicate both in terms of the feature's response mechanism and the feature's perceptible output. In this regard, these feature platforms are disruptive in nature and could be characterized as "game changing" with respect to counterfeit deterrence. They will demand that prospective counterfeiters greatly increase their expertise and investment far beyond that required to replicate optical images. In most cases, the effort to produce features based on these platforms will provide a significant deterrence to all counterfeiting classes. Those counterfeiters with virtually unlimited resources, such as state-sponsored organizations, may have the facilities to replicate such features, but the time required to replicate a specific active response should be considerable. The one caveat to this assessment is that information sharing via the Internet is an emerging game-changing technology that is benefiting the counterfeiter, and this may reduce the advantage offered by these new feature platforms.

Many enhanced security features highlighted in this report are technically sophisticated enough to require instrumentation or, at a minimum, some type of "gadget" or device to aid authentication. These devices could be as simple as a penlight or as readily available as a cellular telephone screen, or more complex, involving machine readers with built-in sensors and logic. This trend toward increased feature complexity is a direct reaction to the reality that, given the advancement in digital and reprographic technology, it is probably not possible to construct a human-perceptible currency feature that cannot be readily counterfeited or simulated.

For anticounterfeiting, future technology appears to be driven by two potentially divergent trends:

1. *The desirability of features that are easily and rapidly evaluated via human senses.* A desirable feature is one for which the assessment is simple and the outcome is Boolean, that is, it may be completely described by a simple "yes" or "no." Features such as "feel" of the substrate and security strips are popular because they can be rapidly sensed and a "yes" or "no" authenticity decision made within the time it takes for the currency note to change hands in a normal transaction.

2. *The increasing ability of counterfeiters to simulate or replicate the desirable features.* This trend motivates the exploitation of specific emerging areas of science and technology for the creation of new features that cannot be readily attempted by counterfeiters. The technology of interest would probably not otherwise be considered for a feature. Indeed, these platforms would only be considered in the context of a proactive strategy of counterfeit deterrence—that is, searching for features that are well beyond the current and projected capability of the most common counterfeiters.

The longer-term features and feature platforms are outlined below. They are grouped into substrate features, composite additions, and electronic platforms. These brief explanations prove a quick look at the new features considered, with additional detail available in Appendix D.

Substrate Features

There is a tremendous potential to develop new features by adding elements to the banknote substrate. The committee suggests two feature platforms that will modify the substrate to significantly improve counterfeit deterrence. One route is to engineer the cotton fiber, or dope the cotton fibers, to impart special properties that would be proprietary to U.S. banknotes. Another approach is to add optical fibers to the substrate in such a way as to form a unique random pattern. Both methods rely on devices for authentication, but they would be highly resistant to counterfeiting attempts.

Digitally Encrypted Substrate

With a digitally encrypted substrate, random patterns can be included in each banknote, and information about that specific pattern can be encoded directly on the banknote for digital authentication. For example, the fiber-infused substrate feature described in Chapter 4 and Appendix C employs optical fibers randomly distributed throughout the substrate. The location of each fiber face could be measured at the point of manufacture, and statistics about that location distribution could be encoded in a bar code printed onto the banknote. In order to authenticate the note, a machine reader would measure the distribution and compare it to the information stored in the encrypted bar code. The potential effectiveness of this feature platform resides in the difficulty of embedding the random pattern throughout the note and requires an unimpeachable algorithm linking the pattern to the bar code.

This feature is based on the idea of adding small-diameter optical-fiber segments to the currency substrate during its manufacture. This process creates a random mixing of the fiber in the paper batch before the paper is dried. The fiber segments, when illuminated by laser light or narrow-spectrum illumination, create a unique optical signature that can be tagged to the specific note. When the finished substrate is illuminated with light, especially laser light, the pattern lights up as the incident light emanates from the ends of the fibers. After the manufacturing cycle of the banknote was nearly completed, a selected region of the note would be scanned or digitally photographed. This image would then be converted to a secure, two-dimensional bar code that would then be printed on the banknote. In order to authenticate the note, a machine reader would compare the image of the selected region to that stored in the encrypted bar code.

In the committee's evaluation, this feature platform has a high rating for counterfeit deterrence owing to the extreme difficulty of exactly duplicating the optical pattern and to the utility of the highly robust image analysis of the fiber placement. Re-creating the exact fiber placement would be virtually impossible. Furthermore, this feature would not be reproducible using electronic printing and scanning techniques, and hence it would frustrate nearly all counterfeiters. It is expected that only a few very persistent counterfeiters would attempt to generate their own substrate with fibers in it. If they did, they still could not duplicate the exact fiber structure in authentic currency substrate materials.

Engineered Cotton Fibers

The cotton fiber is a complex biological structure engineered by both natural selection and intensive human plant breeding. Engineered cotton fibers are not a feature but rather set of tools that may be employed to generate a wide array of potential features. There are a number of ways to engineer a cotton fiber, including the addition of new materials to the fiber lumer, the modification of the cellulosic material that forms 90 percent of the fiber itself, the modification of the proteins associated with the fiber, or the combining of these methods. Engineered cotton fibers offer the opportunity to retain that unique, highly distinctive "feel" of the U.S. dollar, while simultaneously providing a feature that would be extremely challenging to simulate.

Examples of the new capabilities that would be possible range from features that could be detected without any assistance to those that would require a simple device for authentication:

- A highly visible color-shift of the substrate,
- Complex fluorescence patterns incorporated directly in the substrate,
- The filling of the normally hollow cotton fiber with a material in order to tailor the properties of the substrate, and
- A unique spectroscopic signature that the substrate would possess.

Composite Additions

The committee suggests two feature platforms that incorporate advanced materials within the substrate to produce a desired response, tactile or visual, based on the characteristics of the advanced material. One approach, the anomalous currency space (ACS), provides a region or regions within the substrate composed of materials that differ entirely from the banknote substrate itself. The second approach is to incorporate shape memory materials, such as NiTi, in the substrate to exploit its superelastic effect.

Anomalous Currency Space

The anomalous currency space is a materials-based approach that serves as a platform for a wide range of anticounterfeiting strategies by providing a region or regions that differ entirely in materials composition from the banknote substrate. The primary objective is to provide an eye-catching visual feature that would also possess tactile properties. The simplest implementation of this feature platform is a clear plastic window. However, the space does not have to be shaped like a typical window. For example, the shape could be a strip that runs the full length or width of the banknote, or a strip that runs along any or all of the edges of the note, or a series of dispersed regions throughout the currency note. Also, the materials composition of this region does not necessarily need to be a polymer.

Examples of how the ACS provides for the incorporation of heterogeneous materials into the Federal Reserve note (FRN) include the following:

- *A strip along the banknote's outer edge:* A strip of ultratough materials around the outer edge of the bill could be easily detected by its distinct look and feel, such as being resistant to tearing or perforation.
- *Distribution of small ACS features throughout the banknote:* Such patterns could exhibit dynamic as well as passive behavior if, for example, memory metals or polymers were employed.
- *Memory polymers:* A simple case would involve reversible surface "stippling" that would manifest as a change in roughness that could be detected qualitatively by touch and quantified by instrumentation.

NiTi Shape Memory and Superelastic Responsive Materials

Shape memory materials, most commonly based on the intermetallic alloy NiTi, offer various phenomena that can produce active responses which would be useful for developing human-perceptible features. These phenomena are based on the reversible austenite-to-martensite phase transformation. Probably the most exploitable phenomenon is transformation superelasticity, in which structures can be deformed owing to large strains but recover shape when the stress is released. These structures could be wires or thin-foil-based patterns (dots, eagles, numbers, buildings, and so on) that could be easily deformed but would spring back to shape. The NiTi features would provide the public with a dramatic feature that should be robust and easy to identify. Such features would also be very useful for the blind.

The active nature of these features would be a strong deterrent to currency counterfeiting for all but the most sophisticated counterfeiters. The professional criminal counterfeiter would have difficulty duplicating the chemistry (stoichiometry) and processing to achieve a suitable response.

Electronic Platforms

The committee suggests four feature platforms that represent radically new active platforms that produce a sensible response to some form of user input. Described below, they are chemical sensors, e-substrate, smart nanomaterials, and tactilely active piezoelectric features.

Chemical Sensors

Chemical sensors are sensors are embedded in a banknote that detect chemicals generated either by direct human interaction or from a dedicated "pen" and produce a human-perceptible signal. Passive sensors change their appearance directly; active sensors require power that can be either self-generated on the banknote or obtained from a battery at the point-of-sale. Active sensors can trigger visual, audible, or tactile responses (see the subsection below "e-Substrate"). This feature class can enable features for unassisted use or use with devices and features for the blind. The sensed chemical could even be an exhalation gas, activated by breathing on the sensor.

The effectiveness of chemical sensors for banknote authentication depends entirely on how sensitive and robust the sensors are and on the specific implementation of the human-perceptible response. In general, active features (those that change in response to a stimulus) should be easily noticed by the general public. Chemical sensors would be difficult to reproduce by opportunist counterfeiters and petty criminals.

e-Substrate

Two distinct classes of feature platform can be collectively described as electronic (e)-substrates. Both involve a fundamentally new approach that adapts, for the production of currency or currency features, tools and fabrication facilities principally designed for electronics that are used in large-area applications, such as liquid-crystal televisions and computer monitors. The basic approaches used in these proven applications are sufficiently adaptable that they can be implemented with a wide range of substrates (for example, plastics and, in some cases, textiles), materials (for example, polymers) and designs (for example, passive patterns as well as active electronics) that could provide outstanding innovative deterrent features for currency.

The first type of e-substrate feature, referred to here as a passive e-substrate feature, employs the manufacturing techniques used to form patterns, images, and/or multilayer structures on polymer sheets for large-area electronic systems with feature sizes, material types, and overlay registration that lie well outside the capabilities of any existing (or likely future) printing technology. When this type of feature is integrated in a security strip in a banknote paper substrate or, ultimately, when a passive e-substrate is employed as the currency substrate itself, it can provide excellent security. In the second and much more technically challenging type of e-substrate feature, the active e-substrate feature, the patterns that compose the feature are active devices that can respond to a cash handler or a machine reader. Active e-substrate features require three primary elements: power generation, circuitry, and a human-perceptible response. Features such as chemical sensors and tactilely active piezoelectric features described elsewhere in this chapter would use active e-substrate technology.

Smart Nanomaterials

The field of "smart" nanomaterials may provide the foundation for security features that are simultaneously extremely complex to fabricate and easy to use. Smart nanomaterials are currently being developed for a wide range of applications. Many of the smart nanomaterials projected to come out of the nanotechnology revolution will be created by some variation of molecular manufacturing (MM). These materials are expected to exhibit a wide range of dynamic behaviors that may be applicable to anticounterfeiting features. Many research programs in smart nanomaterials have targeted the use of extremely high technology manufacturing systems to create materials capable of independent dynamic responses. The final output of these responses may be both human-perceptible and Boolean, that is, their state represented by a simple "yes" or "no" answer regarding authenticity. These features could also be readily interrogated by sensors. If such a smart nanomaterial can be designed in a manner that fits within the physical and fiscal constraints required for currency, the result would be an anticounterfeiting feature almost impossible to counterfeit or simulate yet as simple to use as color-shifting ink or a watermark.

Smart materials whose macroscopic structure depends on the molecular self-assembly of engineered molecular structures may offer a unique combination of "ultrahigh technology" fabrication that produces simple yet uncounterfeitable Boolean behavior. Molecular self-assembly (MSA) is being explored as a manufacturing base for the large-scale industrial processing of consumer items such as computer chips and scaffolds for bioengineering applications. Large-scale industrial demand for MSA/MM-based facilities may bring the cost of security features using the same technology into line with the economic constraints of FRN production the near future. In a sense, these security features could be considered as a spin-off of the National Nanotechnology Initiative.

Tactilely Active Piezoelectric Features

A piezoelectric crystal develops a voltage when strained. Conversely, if a voltage is applied to a piezoelectric crystal, it responds with a strain resulting in a shape change or deflection. This effect may be used in currency applications to generate a detectable change in the tactile feel at a specific location on a note. In the simplest sense, one can envision bumps that would raise or lower on the note when a voltage is applied. Patterns of fine bumps could possibly be produced to provide each denomination with a unique, readily identifiable pattern. Ideally, the voltage would be supplied by an internal power source embedded within the substrate (see discussion in the subsection above, "e-Substrate"), but the feature could also be powered by a battery that would be attached at point-of-sale locations.

Such features would have a number of advantages for use by the general public. The novel nature of the features would attract attention from the public, leading to the users' taking advantage of the features. Currency users could readily detect changes in the tactile nature of the features. Tactilely active piezoelectric features would be quite effective in deterring counterfeiting by all but the most sophisticated counterfeiter.

SUMMARY AND CONCLUSIONS

This chapter introduces disruptive features that could radically change the currency platform, that is, the banknote. These advancements, if applied to currency features, will hasten the expected shift from printed, static features to advanced substrates and composites with active features. Smart materials and substrates can be used to develop features that require advanced printing methods or other physical processes to create the currency substrate. The e-substrate feature platform and the anomalous currency space are essentially a collection of technologies that can be assembled to produce the desired security capability. The great benefit of these approaches is that they should be highly extensible, with new developments in power sources, printed circuitry, and miniaturized readout devices. These active features create ongoing opportunities for the design and implementation of new currency beyond what is imagined in this report.

A number of these features could be used by the blind through the improved tactile response of banknote currency. For example, the piezoelectric effect used to modify the surface profile of the currency can be used to create surface "bumps" or patterns through some external power source. Currency tactility would be altered with and without electrical power. Also, nearly all of the features can be designed to be readily machine-readable.

Importantly, the features that result from these technologies would present a huge barrier to opportunist counterfeiters, effectively negating the advantages that the counterfeiters have enjoyed from rapidly advanced digital printing technology. In summary, the committee makes the following conclusions:

- The manufacturing processes for features derived from disruptive technologies require special materials or microfabrication techniques, such as nanotechnology, biomaterials, electronics, and advanced materials manufacturing.
- These disruptive feature platforms can provide the basis for a large supply of highly effective features, enabling a proactive strategy of staying well ahead of counterfeiter capability, especially the opportunist counterfeiters who would be ill-equipped to invest in complex, high-technology techniques not dependent on printing.
- Many of the advanced feature platforms require simple tools or devices for authentication. The committee expects that authenticating devices will become pervasive and inexpensive.
- A proactive research, development, and evaluation program would facilitate the identification of highly capable deterrence features. Since these technologies are not mature for security applications, a long-term research and development program would be necessary.

6

A Path Forward

The previous chapters established several important results. First, rapid improvements in digital imaging and reprographic technologies are bringing professional-class capabilities to opportunist and petty criminal counterfeiters. Second, the Internet is affording all classes of counterfeiters improved means of "sharing best practices" and sourcing specialty materials (for example, substrates, color-shifting inks) so that the quality of counterfeits is improving rapidly. The implications of these trends are that the pace of the design cycle of new Federal Reserve notes (FRNs) must be sped up and that new counterfeit-resistant features must be incorporated quickly. The third important result established in this report is that there is an ample supply of concepts for such new features. Potential intermediate-term features (available within 7 years) are described in detail in Appendix C and discussed in Chapter 4. Radical new ideas for fundamentally changing the platform for currency are presented in detail in Appendix D and discussed in Chapter 5. Taken together, these results establish that even though the counterfeiting threat is growing more rapidly, potential means of countering this threat lie on the horizon.

This situation poses a challenge for the Bureau of Engraving and Printing (BEP), however. Faced with the enhanced counterfeiting threats, it will need to respond in new ways to counter them. Over the short term, the bureau will need to find a way to sort through the myriad of possibilities for new features in order to select and to implement those it believes to be most able to counter the threats. Over the longer term, it could establish research partnerships, probably with other federal agencies, to examine fundamental changes to its currency platform. This final chapter applies some of the learning from aerospace product-delivery processes and government acquisition processes to the problem of developing and sourcing these new features and feature platforms by the BEP. While not intended to be comprehensive, this chapter is designed to highlight the magnitude and scope of these tasks in order to indicate at a high level what is involved in accomplishing them.

TECHNOLOGY TRENDS

Advanced features, described in Chapters 4 and 5, cover a wide range of technologies. Anticounterfeiting features on banknotes have in the past been primarily intended as observable features—that is, visible features produced by two-dimensional printing technology. Innovative features rely on a broad set of technologies that can significantly raise the counterfeit-deterrent bar but also require expertise in a range of materials science and engineering disciplines, as demonstrated by Table 6-1, which summarizes the different physical technologies used in the features presented in Chapters 4 and 5. Most of these features apply two or more of these technologies. Today's visible features, by comparison, primarily employ a single physical technology.

The trend in increasing technological complexity is illustrated in Figure 6-1. This figure notionally depicts the temporal evolution of the features that are currently in use and those described in this report. Those new features, discussed in Chapter 4, that can be readily developed for implementation within the next 3 to 4 years employ feature technologies and production methods similar to those used for the current features, even though the new features would be more difficult to simulate and more recognizable. However, the features beyond this generation of innovations will exploit advanced technologies that extend into the composite and active-feature realm and necessarily require new manufacturing processes, such as microfabrication.

The ultimate goal is to move feature technology beyond the capabilities of digital reprographic technology and thereby significantly reduce threats posed by

TABLE 6-1 Technologies Employed in the Innovative Features Described in This Report

Technology Category	Features Using This Technology (no.)
Chemistry	9
Computing	4
Electronics	8
Materials	11
Mechanics	6
Optics	16
Sensors	9

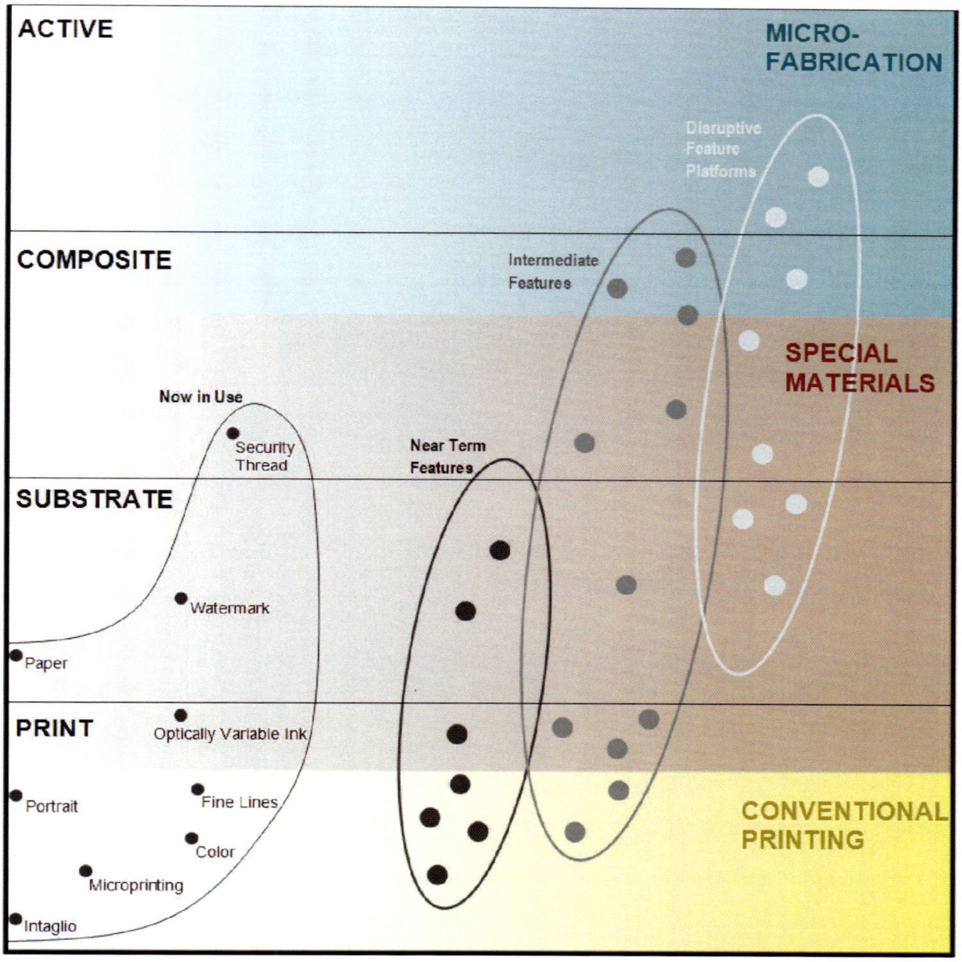

FIGURE 6-1 A notional depiction of the evolution of human-detectable features in currency. This chart shows features that are now in use in U.S. currency and those proposed in this report, as a function of feature technology (left axis)—that is, active, composite, substrate, and print; and as a function of the manufacturing technology (right axis)—that is, microfabrication, special materials, and conventional printing—required to produce the feature. This chart is a schematic that has been drawn to give a general idea of how currency features—and in particular human-detectable features—have evolved and will evolve in the future in relation to technology. NOTE: Near-term features: 3 to 4 years for implementation; intermediate features: within 7 years for implementation; disruptive feature platforms: more than 7 years for implementation.

the opportunist and the petty criminal counterfeiter. The grand challenge is to accomplish this new level of deterrence while maintaining the distinctive character of the U.S. Federal Reserve note and producing these advanced notes at an affordable cost to the Federal Reserve System. This grand challenge is the crux of the research strategy that the committee is hypothesizing for each of the innovative features. The major aspects of this strategy are indicated in the following sections.

FEATURE-DELIVERY PROCESS

The development of innovative counterfeit-deterrent features such as those proposed in this report all the way from the research and development (R&D) stage to implementation in FRNs presents a number of challenges. Not only does the development program of a feature have to meet its expected technical outcome, but also the use of the feature needs to (1) maintain the traditional "look and feel" of U.S. FRNs, (2) allow the FRNs to perform suitably in the appropriate qualification wear-and-tear tests, (3) be compatible with the banknote production process, (4) maintain the effectiveness of the feature over the lifetime of the banknote, and (5) not pose any hazard or use restrictions to the public.

Many of the innovative feature ideas presented in this report leverage work in fields outside of security printing. This work can provide a considerable benefit in accelerating the work required for a full understanding of the physical phenomena that form the basis for the deterrent, as well as providing the technical support for investigating the proof-of-concept.

The considerations highlighted above suggest that a formal phase-gate feature-delivery process is required to create the needed features on budget in a timely fashion. Thus, the committee adapted existing technology- and product-delivery processes in order to propose a four-phase feature-delivery process that could be used by the BEP to convert feature suggestions such as those articulated earlier in this report into delivered new features on FRNs.

Where applicable, references are provided for the features listed to relate what is available in the literature regarding the science and technology to a particular feature or feature platform. Those feature technologies that are well explored have a relatively extensive reference set. There are two additional considerations. The first is that the basic physics of the feature should be obvious. For example, in the case of using grazing-incidence illumination for surface deformations in the currency substrate (the basis for the grazing-incidence optical patterns feature proposed herein), diffraction is the operative physical effect. The idea may be new for currency, but optical science books would adequately describe the physical phenomena.[1] Additional references would not add value in such a case. Second,

[1]Such as M. Born and E. Wolf, 1999, Principles of Optics, Cambridge, United Kingdom: Cambridge University Press; or O.S. Heavens and R.W. Ditchburn, 1991, Insight into Optics, New York: Wiley.

some features may be so new as not to have available references. The major risks of the feature and feature platforms cited in this report are not the fact that their functional capabilities have to be proven, but rather that what has to be proven is that their application in the production of U.S. currency is useful in deterring counterfeiting, that they can be implemented cost-effectively, and that they provide the expected durability and distinctive appearance of the FRN.

Feature-Delivery Phases

The committee regarded the delivery of a new counterfeit-deterrent feature as comprising four phases, which are adapted from NASA's Technology Readiness Levels[2] and the Department of Defense's Manufacturing Readiness Levels.[3] Phase 0 begins the process with idea generation, a thorough literature review, the identification of related work, and an outline of a plan for Phase I. Phase 0 concludes with a specific list of what should be demonstrated in order to make the feature a viable candidate for insertion in a banknote. Phase I focuses on obtaining proof-of-concept for the new feature idea. During this phase, proof-of-concept analyses and experiments are conducted, and application scenarios are developed. During Phase II, feature attributes are developed, along with the requisite manufacturing technology, so that the feature can meet the requirements for a targeted banknote application. Phase III begins with a decision to incorporate the new feature in a specific banknote. At the completion of Phase III, the feature is ready for full production in a banknote.

At the conclusion of each of these phases, a decision would be made to terminate or to continue the development of the feature into the next phase. Of course, the R&D effort can be halted during any phase if the results or projected benefits are not attractive enough for eventual implementation. A description of the four individual phases is summarized in Boxes 6-1 through 6-4:

- Phase 0: Idea Generation and Scoping (Box 6-1),
- Phase I: Establishment of Feasibility (Box 6-2),
- Phase II: Demonstration of Applicability (Box 6-3), and
- Phase III: Maturation and Scale-Up (Box 6-4).

---

[2]The definition of Technology Readiness Levels is available at the Web site of the National Aeronautics and Space Administration (NASA) at <http://esto.nasa.gov/files/TRL_definitions.pdf>. Accessed February 2007.

[3]The definition of Manufacturing Readiness Levels is available at the Web site of the Defense Acquisition University at <https://acc.dau.mil/CommunityBrowser.aspx?id=18231>. Accessed February 2007.

BOX 6-1
Phase 0: Idea Generation and Scoping

Objective: Provide specificity to a counterfeit-deterrent feature idea. Accomplishing this will require the following:

- Performing a comprehensive literature review.
- Identifying related work, including that within universities, industry, federal agencies, and national laboratories.
- Describing expected benefits, key development challenges, and major milestones for further development.
- Outlining a plan for Phase I, which could include partnering with a federal agency with relevant expertise.

BOX 6-2
Phase I: Establishment of Feasibility

Objectives: (1) Demonstrate the feasibility of a new feature and (2) identify credible application scenarios. Reaching these objectives would require the following:

- Formulating and refining the technology concept.
- Identifying candidate counterfeit-deterrence applications for the feature technology.
- Evaluating multiple design options for the application of the new feature in a current Federal Reserve note.[1]
- Conducting adversarial analysis of the proposed feature, including analysis of the ease of simulating the feature and of the degree of detectability by the intended audience.
- Developing a list of the key requirements, as quantitative as possible, that the new feature would have to meet to be considered for a future banknote application.
- Identifying the key challenges that must be addressed to achieve the requirements.
- Establishing the initial feasibility of a new feature technology, including analytical analyses and proof-of-concept experiments.
- Conducting a preliminary assessment of the manufacturing concepts and producibility needs based on laboratory studies and estimates of the implementation cost.

[1]There is a clear understanding that the new feature would not be used for this application, but the exercise would be important to calibrate the benefit of the feature based on a known application.

BOX 6-3
Phase II: Demonstration of Applicability

Objectives: (1) Identify a specific targeted application that would significantly benefit from the new feature and (2) obtain sufficient data to characterize the risks, costs, and benefits so that an "application go-ahead" decision can be made. Reaching these objectives would require the following:

- Demonstrating the capabilities of specific instances of the feature concept—focusing on those options with most significant benefit and realistic implementation costs.
- Iterating the feature technology to optimize the most critical properties and checking the other key material parameters to ensure that these would not pose unacceptable constraints.
- Completing a design of a test banknote incorporating the new feature in order to validate the requirements for the feature and then producing a limited number of these notes for testing, including assessing durability, recognizability, and difficulty in simulating.
- Identifying the investments required to scale up the production of the new feature; developing and instituting the processes required to ensure the producibility and quality of the new feature; fully understanding the manufacturing cost drivers; conducting producibility assessments.
- Demonstrating and validating the new feature in a prototype banknote that was produced using production-scale processes (as opposed to prototype processes) and suitably testing it under relevant conditions.
- Defining and validating manufacturing requirements; demonstrating materials, machines, and tooling, personnel skills, and quality-control methods in a relevant environment; developing or nearly developing key manufacturing processes and procedures.
- Identifying and understanding long-lead items and key supply-chain elements; initiating the production scale-up of the new feature.

Some features could be expressly designed for detection by commercially available currency-reading equipment. Currently the BEP does not provide banknote feature specifications to commercial equipment manufacturers. Registered vendors have an opportunity to examine "test decks" containing relevant features prior to the issuance of a new series note. They use these decks to determine which feature(s) to select for detection and what technical approach to employ. The resulting equipment is then marketed to vending-machine suppliers, retail stores, commercial banks, and so on. For the future, as machine reading becomes more pervasive and the feature technology more sophisticated, the government may consider a proactive approach in providing these registered vendors with test decks earlier, possibly during Phase II of the development program. This approach would allow the vendors more time to develop and optimize their detection technologies for the new features. This capability would provide a useful benefit to the public

BOX 6-4
Phase III: Maturation and Scale-Up

Objectives: (1) Incorporate the new feature into the design of a specific banknote series, (2) qualify the new feature, and (3) complete efforts necessary to scale up the feature production and incorporate it into the banknote. Achieving these objectives requires the following:

- Establishing the processing details of the new feature and issuing the necessary specifications and standards.
- Completing producibility risk assessments and ensuring that production cost estimates meet goals and that the forecasted production schedule is attainable.
- Completing the production scale-up of the feature and fully establishing the supply chain.
- Generating a complete set of necessary feature property data, using features produced via the full-scale production process.
- Fully demonstrating a prototype banknote with the new feature in a realistic environment.
- Finalizing the design of the targeted banknote using the complete feature property data set.
- Qualifying the new banknote through test and demonstration.
- Ensuring that all materials for the new feature are in production and available to meet schedule and manufacturing and that quality processes have been proven and are under control.
- Launching the new banknote.

and commercial cash handlers who expect machine readers to work effectively with new series notes, as well as with the older series notes. Necessary precautions would have to be taken to ensure that the earlier consultation process with vendors did not threaten the security of new features.

Development Risk and Issues

For each feature discussed in Chapter 4, the committee identified a list of significant risks and issues that should be addressed during a development program for that feature. These risks and issues (presented in Appendix C) include an understanding of the feature's inherent durability limits; appearance and aesthetic considerations; issues related to social acceptability, such as potential health hazard concerns and loss of privacy—for example, the ability to scan a person and determine the amount of cash being carried; and key technical challenges.

Table 6-2 summarizes the key technical challenges, abstracted from Appendix C, for each feature described in Chapter 4. Similarly, Table 6-3 summarizes the

TABLE 6-2 Key Technical Challenges for Intermediate-Term Innovative Feature Concepts Described in Appendix C

Innovative Feature Concept	Key Technical Challenges
Color image saturation	—Ensuring that embedding the watermark in the saturation channel does not noticeably degrade the quality of the image. —Demonstrating that the encoded watermark data are altered if the image is copied. —Developing a simple, cost-effective device that accurately detects the saturation watermark. —Evaluating how the watermark is damaged or compromised when copied in a manner that is generally available to counterfeiters.
Fiber-infused substrate	—Ensuring durability of the fiber in the substrate. —Providing for visibility of the fibers in reflected and transmitted light.
Fresnel lens for microprinting self-authentication	—Affordably manufacturing a thin Fresnel lens, on the order of 10 microns thick. —Reliably attaching the Fresnel lens across a hole in the banknote without deteriorating the durability of the note.
Grazing-incidence optical patterns	—Developing appropriate patterns and patterns that are compatible with tolerable analysis procedures.
High-complexity spatial patterns	—Determining the differential quality of the intaglio spatial bandwidth and that of current and expected digital printers.
Hybrid diffractive optically variable devices	—Developing optimized designs that have suitable lifetime and acceptable manufacturing cost.
Metameric ink patterns	—Developing ink that maximally stresses many, if not most, electronic printers. —Designing the ink to be capable of being visually identifiable with simple light sources such as a portable penlight, light-emitting diode (LED) light, and so on.
Microperforated substrate	—Integrating a new processing step into the Bureau of Engraving and Printing's (BEP's) manufacturing process.
Nanocrystal pigments	—Demonstrating ink formulations that exploit the unique optical properties of nanocrystals. —Demonstrating low-cost printing capability. —Demonstrating high durability.
Nanoprint	—Designing a nanoprint feature that would be visible without a device. —Ensuring producibility at acceptable cost on a plastic substrate that is subsequently integrated into the paper substrate with high durability.
Refractive microoptic arrays	—Conducting laboratory demonstration of an outstanding three-dimensional visual and tactile effect. —Providing for durable integration of paper and plastic films, as well as maintaining visibility during use. —Ensuring producibility at low cost.
See-through registration feature	—Ensuring precise alignment, or registration, of the printed feature on the front and back sides of the banknote.

continued

TABLE 6-2 Continued

Innovative Feature Concept	Key Technical Challenges
Subwavelength optical devices	—Determining if embossing-based fabrication approaches, similar to those used for diffractive optical devices, can be scaled up for low-cost manufacturing of subwavelength optical devices.
Tactile variant substrate	—Developing a reproducible process to control substrate roughness. —Understanding how ink will adhere to a substrate with varying levels of roughness.
Thermoresponsive optically variable devices	—Determining low-cost, high-volume manufacturing approaches for thermoresponsive optically variable devices. —Assessing the levels of reliability and durability for currency applications and determining if suitable packaging systems can be developed.
Window	—Developing the window design. —Ensuring the durability of the window.

key technical challenges, abstracted from Appendix D, for each feature described in Chapter 5. The committee envisions that these challenges would be refined during Phase 0 and would become the focus of development activity during Phase I in order to establish the feasibility of the feature concept.

The committee suggests that every feature idea, regardless of the amount of related prior work, begin development with a Phase 0 task. This would establish a firm baseline for the development effort. For features that employ the more mature technologies, Phase 0 could be relatively short. Similarly, the effort to demonstrate proof-of-concept during Phase I may not be large for these features. It is critical though that feasibility be demonstrated against banknote design requirements before proceeding to Phase II. By contrast, immature feature concepts could require considerable effort in Phases 0 and I.

As one might expect, a good anticounterfeiting feature should be aesthetically pleasing, since currency is not just functional but also a form of national identity that is especially recognizable. Thus, the features suggested in this report have been examined with respect to their effect on the FRN's overall aesthetics. Clearly, aesthetic design is difficult to quantify, but any new features should not adversely impact the look and feel of the currency. For many of the features proposed in this report, gauging the aesthetic impact will not be possible during Phase I because the final form of the feature has not yet been determined. However, the committee considered what potential impact a proposed feature could have on the look, feel, and overall appearance of an FRN. Concerns regarding the aesthetics of a particular feature are listed under challenges and should be evaluated along with the other challenges.

Table 6-4 summarizes the key milestones that should be achieved during Phase I for the features discussed in Chapter 4. Specific milestones for Phases II

TABLE 6-3 Key Technical Challenges for Long-Term Innovative Feature Concepts Described in Appendix D

Innovative Feature Concept	Key Technical Challenges
Anomalous currency space (ACS)	—Determining the design concepts to focus on. —Selecting the material system(s) of interest. —Physically incorporating the ACS into the Federal Reserve note.
Chemical sensors	—Selecting the human-produced chemical to detect that covers the range of human variability. —Ensuring the long-term durability of the sensor and its power source. —Ensuring the accuracy of the sensor over the range of operating conditions and multiple sensing attempts. —Establishing with extremely high confidence that the sensor will not collect and transmit human pathogens. —Controlling the cost of implementing the sensing system.
Digitally encrypted substrate	—Selecting the optical fiber that possesses necessary characteristics for incorporation in the substrate. —Selecting the appropriate three-dimensional random pattern and encrypting the pattern in such a way that an authenticator can be printed directly on the banknote. —Addressing the ability of the fibers to survive the high-pressure intaglio process without unacceptable breakage. —Ensuring the durability of the pattern, and the printed encryption of the pattern. —Implementing a method to quickly read the pattern and compare the results with the printed authenticator. —Controlling the cost of implementing the feature.
Engineered cotton fibers	—Determining if phenotypes created in a cotton fiber by rDNA technology will be as durable as those displayed by current fibers. —Evaluating the environmental consequences of growing engineered crop plants, including preventing unintended release of novel plant genes into the environment.
e-Substrate	• Passive structures: —Determining pattern layouts and materials sets that offer maximum benefit. —Developing durable means to integrate with the paper currency substrate, plastic security strips that incorporate these layouts and materials. • Active structures: —Determining the power-generation and packaging requirements for electronics and sensors. —Demonstrating sufficient power generation either on-note or by readily available battery source. —Demonstrating reliable operation of electronics, power source or connectors, and human-perceptible response.
NiTi shape memory and superelastic responsive materials	—Developing cost-effective processing with the necessary chemistry control, including foil processing and physical vapor deposition.

continued

TABLE 6-3 Continued

Innovative Feature Concept	Key Technical Challenges
Smart nanomaterials	—Selecting materials of interest. —Understanding the risks associated with environmental, health, and safety effects of nanotechnology.
Tactilely active electronic features	—Determining the durability of tactilely active piezoelectric features. —Demonstrating scale-up of piezoelectric patterns that are integrated with an electronic substrate. —Demonstrating an effective power source for the devices, either external or integrated within the substrate.

and III, beyond what is described above in general, depend to a large extent on the results of Phase I. As previously mentioned, some of the innovative features in Table 6-4 will proceed through Phase I rather quickly owing to their simplicity, the extent of related development work, or their potential to serve as a direct replacement for an existing feature. Other features will take considerably longer for a demonstration of proof-of-concept and feasibility.

All of the long-term feature platforms summarized in Table 6-3 will require considerable time for development. The primary Phase 0 milestone for these features would be a roadmap for the most promising development routes. One strategy for development of these long-term concepts is to maintain a long-term core development effort in Phase I. As the Phase I work proceeds, specific feature ideas would be identified for a targeted application. In this case, a new project would be started in Phase I for the feature idea that was spun out of the more general development program for the feature platform with the intent of proceeding on to Phases II and III as the technology is matured. For example, the e-substrate is not a feature itself but an enabler for other active features. The Phase I effort would maintain awareness of related developments and would conduct feasibility assessments of the most promising approaches. If an active feature were identified as a high priority, the appropriate elements of the e-substrate would be incorporated into the active-feature development program as it progressed from Phase I into Phases II and III. By contrast, a specific feature idea, such as engineered cotton fibers, could proceed through the development phases once feasibility was established in Phase I.

Single Feature Versus a Set of Features

The committee acknowledges one key limitation of this report: the innovative features and concepts were evaluated by considering the effectiveness of each fea-

TABLE 6-4 Summary of Phase I Milestones and Timing for Intermediate-Term Innovative Feature Concepts Described in Appendix C

Innovative Feature Concept	Key Milestones—Phase I	Estimated Time to Complete Phase I
Color image saturation	—Selecting an image or images suitable for use on currency. —Saturation watermarking the images, scanning and/or copying them, and processing the data. —Investigating how well the saturation watermark passes through the currency engraving and generation process. —Evaluating how the watermark is damaged when copied.	●
Fiber-infused substrate	—Placing fiber fragments in paper substrates to determine the applicability of the technique and any operational or manufacturing difficulties that might arise. —Scanning and visually observing the fragments according to the referenced techniques to see how difficult practical implementation and verification of this feature might be. —Developing prototype scanners and associated processing software.	●●
Frensel lens for microprinting self-authentication	—Demonstrating the fabrication of a Fresnel lens with thickness $\leq$10 microns, aperture $\geq$10 mm, and focal length ~25 mm, operating in the visible spectrum. —Demonstrating an approach for integrating the lens over a hole in the substrate such that it passes all durability tests. —Defining a lens fabrication process that has the potential to be affordably scaled up.	●●
Grazing-incidence optical patterns	—Designing test impressions for implementation and evaluation. —Generating impressions in substrate material using equipment that would be used for production of the substrate or note. —Developing evaluation systems for analysis of the feature. —Designing production feature setup and testing.	●
High-complexity spatial patterns	—Carrying out image development and evaluation of targets containing the desired spatial patterns. —Rendering the desired target in a form that can be printed with the intaglio process on substrates of interest. —Attempting to reproduce the pattern using current high-quality digital copying and reproduction methods.	●
Hybrid diffractive optically variable devices	—Establishing preliminary designs and layouts and testing them in the currency substrate. —Performing durability tests to study the effects of wear.	●●
Metameric ink patterns	—Developing a metameric ink that meets cost and durability targets, as well as demonstrating adequate metamerism in the expected user environments. —Developing and/or identifying simple light sources such as penlights with filters or light-emitting diode penlights that could be used at point-of-sale locations to assess the authenticity of the currency quickly. —Evaluating the ink's applicability to printing equipment at the Bureau of Engraving and Printing.	●

continued

TABLE 6-4 Continued

Innovative Feature Concept	Key Milestones—Phase I	Estimated Time to Complete Phase I
Microperforated substrate	—Assessing durability. —Demonstrating effectiveness.	●
Nanocrystal pigments	—Addressing the toxicity issues related to metallic and semiconductor crystals; down-selecting materials that are clearly nontoxic. —Carrying out laboratory-scale formulation of an ink from nanocrystals. —Carrying out proof-of-concept of unique color and spectral characteristics.	●●
Nanoprint	—Developing a preliminary nanoprint feature that is visible without a device. —Exploring technical approaches to producing nanoprint features at acceptable cost on a controlled plastic substrate. —Exploring methods to integrate a plastic substrate into the paper substrate with high durability.	●●
Refractive microoptic arrays	—Defining and showing initial feasibility for printing optical elements at low cost. This would include a high-resolution stamping process for optical elements of this size or embossed films; determining if sufficient resolution exists to print simultaneously on reverse side to attain required registry, and exploring a viable approach for printing the elements on a high-speed drum printing flexo/offset press. —Exploring approaches to add color three-dimensional perception to the optical effect. This would include a pigment and dye approach, or alternately an interference filter stack, high printing resolution, and integration of advanced dye or pigment with high-resolution offset printing. —Developing robust integration of a microoptical device with the paper substrate, including laminated plastic film optical elements to achieve required durability. —Exploring the creation of varying sizes, shapes, and positions of the window. —Investigating an increase in the tactile effect.	●●
See-through registration feature	—Designing a test see-through registration feature that is both pleasing to the eye and easy to authenticate. —Testing existing BEP printing equipment for precise registration.	●
Subwavelength optical devices	—Developing a reasonable approach to achieve high-volume production of subwavelength optical devices for operation in the visible range. —Developing prototype scanners and associated processing software to detect subwavelength optical devices.	●●●

TABLE 6-4 Continued

Innovative Feature Concept	Key Milestones—Phase I	Estimated Time to Complete Phase I
Tactile variant substrate	—Conducting a laboratory demonstration of a substrate-roughening method that can produce roughened areas of different shapes. —Developing test features incorporating tactile variance. —Conducting experiments to estimate the durability of a roughened substrate. —Carrying out initial experiments to determine how intaglio images will be affected by a substrate with different levels of roughness.	●
Thermoresponsive optically variable devices	—Conducting initial durability tests of several different thermoresponsive optically variable devices. —Developing a reasonable approach to achieve cost-effective integration of thermoresponsive optically variable devices in the currency substrate.	●●
Window	—Solving any durability issues. —Finalizing whether the windows would be best produced prior to or after printing. —Finalizing the window-producing process.	●

NOTE: "●" indicates that the estimated time to completing Phase I is within 2 years; "●●," 2 to 3 years; "●●●," 3 to 4 years.

ture independent of all other features that might be present in a singe banknote. To the degree that the committee has studied the interaction of features, it can, however, conclude the following:

- Many features are not truly independent—they can complement, or in some cases degrade or even negate, the effectiveness of other features.
- Reliance on a single dominant feature can provide false security if that feature can be replicated well enough and users do not inspect other features. Prominent features can create a focus of attention that negates others, so that the counterfeiter can replicate or simulate the dominant feature and ignore other security measures. For instance, color is a dominant feature. But current reprographic technology can easily replicate the appearance of the color note, thereby decreasing the chance that a consumer would check other features. Thus, avoiding over-reliance on a dominant feature should be included in the evaluation process for selecting features; this consideration would also apply to machine-readable features.
- To the degree that several new features may be more effective than a single feature, they should be incorporated as a set of features so as to provide

a layered defense against an array of counterfeiting threats. Such an approach will create a public awareness of multiple features on the note and reduce the reliance on a single dominant feature. Multiple features can also address the various classes of counterfeit threats; for example, color image saturation and metameric ink features would target nonprofessional counterfeiters, while the fiber-infused substrate and the nanoprint features would target professional counterfeiters.

BANKNOTE DESIGN PROCESS

The feature ideas that originated during the course of this study arose out of a technology-push perspective—that is, the committee considered what could be possible and then used various criteria to refine the list. This report necessarily does not contain an exhaustive list of all possible new features. Additional technologies undoubtedly could be successfully employed to improve counterfeit deterrence. However, the committee believes that the process followed in conducting this study would be applicable to the analysis and development of other candidate features. Ultimately, in order to be useful, an innovative technology must be proven effective. The goal is to provide banknote designers with new counterfeit-deterrence feature options that they can employ to stay a step or two ahead of the advancing counterfeit technology. These options will allow the designers to engage in assessing the "art of the possible" as they (1) synthesize the totality of needs and requirements that a new banknote must satisfy, (2) analyze the functions that the new banknote must perform in order to meet those requirements, and (3) determine how best to allocate requirements to the various components and features that comprise the banknote. The collection of these activities required to design new features and incorporate them into banknotes is the banknote design process.

This proactive approach differs significantly from the traditional approach, which entails an evaluation of mature counterfeit-deterrence features for which most of the issues have already been addressed, for which a supply chain exists, and which are already in use with known effectiveness.

For the proactive approach, candidate innovative features would proceed through the development phases with an increasing focus on a targeted banknote application. Therefore, the advocacy for feature development must transition to a technology-pull mode in which the banknote designer has a significant role. The successful features will survive trade-offs between competing factors with respect to the banknote's production cost, implementation schedule, and technical performance. Also, during the phased development process, some of the innovative features will be found to be noncontenders for insertion. The reasons would include shortfalls in performance that cannot readily be overcome, a high production cost that is not amenable to cost reduction during scale-up, excessive time to complete

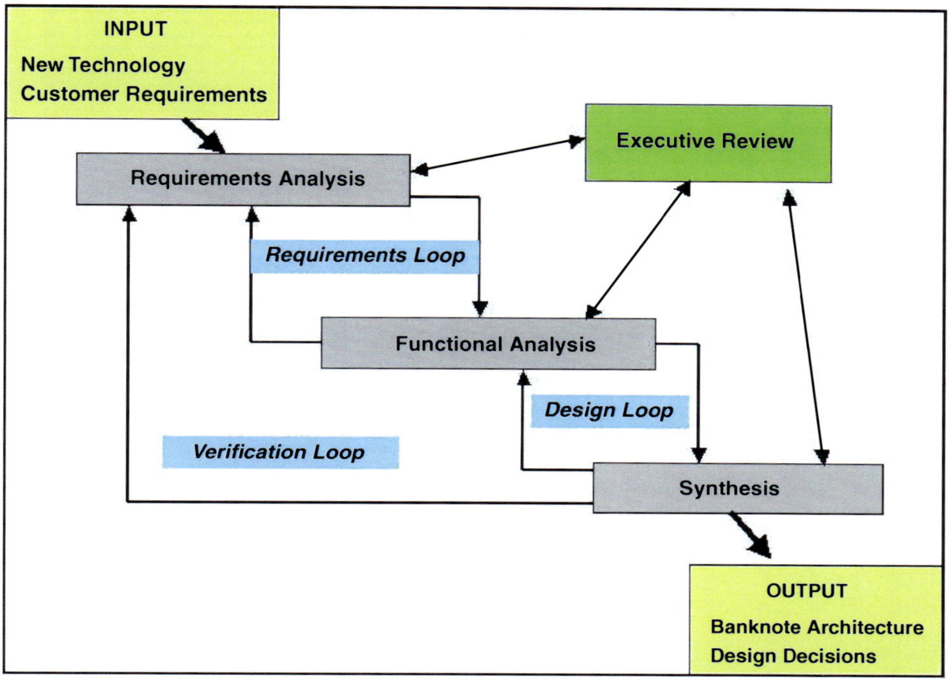

FIGURE 6-2 Design engineering process. SOURCE: Adapted from John B. Wissler, 2006, Technology transition: A more complete picture, Defense Acquisition Review Journal 13(1):10 (Fort Belvoir, Va.: Defense Acquisition University Press).

development because of the large number of complex issues that must be addressed, and the ready availability of another feature that can better satisfy almost all of the key requirements.

The design engineering process is necessarily an iterative one, as depicted in Figure 6-2. It is through this mechanism that the transition from the technology-push to the technology-pull development focus occurs. The design process is overseen by an *executive review* that allocates resources, establishes time lines, conducts trade-offs between suites of features, and makes "go/no-go" decisions on the continuation of a particular project. As illustrated in the figure, the key starting point of the design engineering process is the *input* of new technology that occurs during Phase 0. *Requirements* definition is one of the primary activities initiated during Phase I. Through this mechanism, a banknote designer can establish the requirements for a feature and then evaluate the proof-of-concept results against these requirements. As the development proceeds through the subsequent phases, these requirements will be refined and developed in more detail until they are finalized during Phase III, as indicated by the Requirements Loop. *Design* of a feature is also an iterative process. It is formally initiated during Phase II when the prospective feature is targeted for a specific application. This serves as the basis for identifying needed information, so that the effort expended during Phase II is focused on collecting the data to support these information needs. Further data will be collected during Phase III to support the production go-ahead decision as functional analysis and synthesis of design solutions are accomplished. Finally, *verification* tests the design solution against the latest requirements. This is a key activity as Phase III proceeds. If the design solution falls short of requirements, further review of the requirements and the design solution would be necessary, as depicted by the Verification Loop. When Phase III is successfully completed, the new feature will be incorporated into the banknote architecture, ready for production.

PROGRAM-EXECUTION CONSIDERATIONS

Many more feature ideas are presented in this report than can be funded for development. The committee was unable to further prioritize the features beyond selecting the most promising ones for listing in the report. The experts at the BEP could use the information in this report as the basis for further prioritization, thereby determining which features have the most promise for addressing gaps in deterrence beyond what is already planned. It would be important to consider the long-term as well as shorter-term needs to establish a balanced portfolio of features for development.

Some of the feature ideas in this report could readily be explored by the BEP working with equipment and other suppliers. Other feature concepts will require considerable expertise in technology areas for which the BEP has had little, if any, experience. In these cases, partnering with another federal agency that already has relevant expertise could allow the BEP to make rapid progress without incurring large expense if it is the intent of the bureau to own the intellectual property rights of the new feature(s). At some point in the development program, such as during Phase II, the supporting supply chain would be incorporated into the program.

The committee has identified a number of new ideas that require further exploration, and other ideas can be added. These ideas are extremely broad ranging, from an idea that could readily proceed through the development stages in a few years (for example, a see-through registration feature) to game-changing concepts that require significant development over multiple years, perhaps a decade (for example, smart nanomaterials).

Although it is outside the committee's charter to recommend a particular scope for such a research program, the magnitude of the task of evaluating the concepts articulated in Appendixes C and D suggests the need for a two-pronged approach. For the intermediate-term innovative features, the essence of the task is to select the most promising features for immediate implementation. The BEP could perform this task internally. Although suppliers can help with certain details, the research to evaluate and to select certain features above others could be done internally.

For the longer-term, disruptive platforms, partnerships with other federal agencies seem to be an attractive option. In this case, the BEP would work with other agencies to get its objectives embedded into long-term joint research programs on advanced technology (for example, nanotechnology). The currency experts at the BEP would review the results of these programs jointly with the primary sponsoring agencies on a regular basis to ensure that its needs are being addressed adequately. By analogy with comparable government and industry efforts, the committee believes that a significant multimillion-dollar annual budget would be required to provide seed funding for Phase 0 and Phase I projects, with the amount determined by BEP requirements. A program of this scale would allow a balanced program of low-risk and high-risk, game-changing technologies to be pursued. This could build a portfolio of counterfeit-resistant features addressing short-term, intermediate-term, and long-term threats. Once plans were solidified and priorities established in Phase I, appropriate funding for Phase II and Phase III efforts could be established.

FIELD TESTING OF FEATURES

During the course of this study, the committee observed that there does not appear to be a scientifically based federal program dedicated to field testing the effectiveness of existing and proposed banknote features and feature sets. It is normal commercial practice to conduct thorough market tests of new products. The committee was made aware of only two studies of the effectiveness of some features used on older series notes.[4,5] These laboratory-based experiments presented suggestions concerning the various factors that might influence the "passing effectiveness" of a note, but they do not produce the quantitative information needed to estimate the chance that a particular banknote feature combination would be successfully passed in an actual, real-world transaction.

Several major banks and commercial currency manufacturers have active programs that study feature effectiveness and pursue research on new technologies to deter counterfeiting. These efforts often supplement aggressive public awareness campaigns by a central bank to introduce new security features in banknotes. Also,

[4]A.P. Hillstrom and I.H. Bernstein. 2002. Counterfeit detection for new and old currency designs. Proceedings of SPIE, Optical Security and Counterfeit Deterrence Techniques IV, R.L. van Renesse (ed.), Vol. 4677, pp. 65-80.

[5]R.M. Klein, S. Gadbois, and J.J. Christie. 2004. Perception and detection of counterfeit currency in Canada: Note quality, training, and security features. Proceedings of SPIE, Optical Security and Counterfeit Deterrence Techniques V, R.L. van Renesse (ed.), Vol. 5319, pp. 1-12.

the committee noted that the Bank of Canada—the Canadian central bank and equivalent of the Federal Reserve—conducts research in the visual and optical evaluation of banknote features, including user studies to test the ability of professionals and nonprofessionals to detect counterfeit notes. The National Printing Bureau of Japan, which prints Japanese currency, includes on its staff expertise in image processing and optics, and its researchers regularly publish in conferences related to digital imaging and counterfeit technology. De La Rue International (with Portals Bathford), a private company that prints currency for a multiplicity of countries, employs expertise in image processing and optics in its feature planning and production. Many international currency-related organizations employ, on a continuing basis, teams of experts similar to the committee that produced this report, and their research extends to the sponsorship of collaborative research with industry and academia partners.

The BEP could pursue a similar, ongoing research activity to conduct formal, statistically valid field tests to determine the effectiveness of features, materials, and technologies in banknote currency. Currently, the U.S. Secret Service collects data on counterfeit usage domestically and abroad, and the BEP has several effective measures to test the durability of banknotes. These efforts effectively monitor counterfeit use and sustainability. They are not designed to address the effectiveness and social acceptability of security features. Research that evaluates the visual, tactile, and aesthetic acceptance of banknote features and the ability of the different types of human cash handlers to use these features to detect counterfeit notes could provide valuable insight into future banknote design.

CONCLUSIONS

- Rapid advances in digital imaging and reprographic technologies have created enhanced threats of counterfeiting which, in turn, could compel the Bureau of Engraving and Printing to respond in new ways to the evolving threat.
- A proactive strategy of incorporating counterfeit-deterrent features that are a step or two ahead of the technology available to counterfeiters requires that government efforts be directed at feature innovation rather than focused only on feature integration.
- The committee has identified a list of innovative features that have a high potential to deter counterfeiting, particularly by opportunist and petty criminal counterfeiters. In general, these features employ technologies that are not being developed for currency applications and will not advance to a stage at which they can be readily incorporated into Federal Reserve notes without dedicated research and development for application.

- In the future, as machine reading becomes more pervasive and banknote feature technology more sophisticated, the government may consider a proactive approach in working with equipment and other suppliers during the development of new features. This would benefit the public that expects machine readers to work effectively with new series and older series notes.
- There is a clear need for sustained research and development of banknote features, materials, and technology. The innovative feature ideas presented in this report require R&D funding to establish feasibility for application in U.S. banknotes and to make them ready for production implementation. By analogy with comparable government and industry efforts, the committee believes that a significant, multimillion-dollar annual budget would be required to provide seed funding for Phase 0 and Phase I projects of low- and high-risk technologies, with the amount of funding determined by BEP requirements.

Appendixes

A

Statement of Task and Organization of the Report

STATEMENT OF TASK

Primary Tasks:

1. Identify technologies, both existing and emerging, that pose the most significant counterfeiting threats to Federal Reserve notes (FRNs). Threats known today include digital methods of producing images, desktop scanners, digital cameras, color printers, digital imaging software, and digital pre-press and printing equipment. The evaluation should include existing emerging threats to FRN features used by the general public to authenticate currency, as well as features used in vending, ATMs, retail sorters, the gaming industry and other automated currency processing.
2. Identify features, materials, and technologies to deter counterfeiting of FRNs, and assess their relative effectiveness. The study should include the identification, analysis, evaluation, and ranking by effectiveness of technologies that may deter the counterfeiting of FRNs and that could be incorporated into U.S. banknotes in the longer term (more than 5 years). The evaluations of technologies should include the following criteria:
   a. Effectiveness in deterring the counterfeiting of FRNs (i.e., difficulty in duplicating or simulating FRNs using existing or emerging commercially available materials and processes).
   b. Promoting visual authentication (i.e., technologies that are visually distinctive and obvious to the untrained observer, as well as noticeable,

understandable, and easily used by the general public as a method of visually authenticating FRNs in a variety of lighting conditions).
   c. Uniqueness and aesthetics (i.e., novel or strikingly different from existing features used to deter counterfeiting of high-security documents, and aesthetically pleasing in the design of FRNs).
3. Identify potential costs, including material costs, equipment costs, and the costs of processing banknotes for the Federal Reserve System and third-party users of FRNs, including transportation, storage/handling, and eventual disposal. Feature evaluation should evaluate the implications of implementing proposed materials, technologies, or features on the BEP's FRN manufacturing operations, including the following:
   a. Evaluation of the merits of exploiting the three-dimensional character of a banknote. Development of a new class of deterrents based on compositional changes of the substrate [the surface or material on which printing is done], incorporating new materials in a variety of new innovative ways, or incorporating optical or auditory security elements into the substrate.
   b. Evaluation of alternative banknote substrates relative to each other, and the potential of blending various substrates with other substrates—including standard banknote paper to create a hybrid that expands value as a counterfeit deterrent or new security feature. Include substrates already used for banknotes worldwide, as well as potential materials not yet in use, but that may have significant potential benefits.

Secondary Task (not required, but to be completed if time and funds allow):

4. Identify, including analysis, evaluation and ranking of the effectiveness of technologies that could be incorporated into FRNs in the long term (more than 5 years) for denominating or authenticating by the blind, for forensic analysis, and for third-party machine denominating or authenticating (e.g., point of sale), including estimated costs of implementation.

ORGANIZATION OF THE REPORT

Chapter 1, "Background and Motivation for the Study," provides background information and an overview of the main motivations for the report and is an important foundation for understanding the directions taken in the study by the committee. Chapter 2, "Understanding Counterfeiting," describes five classes of counterfeiters and summarizes a flow model of counterfeiting that can be further developed to provide quantitative insight into the deterrence effectiveness of U.S. banknote features. While not answering any one element of the charge—although it is most pertinent to Task 1 of the statement of task—Chapter 2 also provides important background information about the threat to currency and how the committee established a system to evaluate potential new features and feature platforms. The committee found that much of the information in this chapter was not at all obvious and as such was valuable to share with the wider audience for this report.

Chapter 3, "Emerging Counterfeiting Technology Threats," which addresses Task 1, reviews current and emerging technologies that present significant counterfeiting threats. Chapter 4, "Innovative Counterfeit-Deterrent Features" discusses the evaluation process used by the committee and summarizes the advanced features that could be incorporated in FRNs within a short to intermediate time frame—that is, 4 to 7 years. Features particularly useful for classes of users, such as the blind, are identified, and plastic substrates are discussed. The features summarized in Chapter 4 are described in more detail in Appendix C. Together, Chapter 4 and Appendix C address a portion of the feature identification and prioritization undertaking requested in Tasks 2 and 4 and the alternative substrate evaluation requested in Task 3.

Chapter 5, "Disruptive Feature Platforms," presents game-changing ideas for feature technologies that extend beyond conventional approaches and which will require long-term development (beyond 7 years). Appendix D contains more detailed descriptions of the platforms. Chapter 5 and Appendix D address a portion of the feature identification and prioritization requested by Tasks 2 and 4 and offers long-term concepts for alternative substrates, as requested in Task 3.

Chapter 6, "A Path Forward," discusses a strategy for further developing the feature concepts to the point that a decision could be made for their incorporation into an FRN. Factors of cost, aesthetics, and durability are explained. Chapter 6 addresses the remaining elements of Tasks 2, 3, and 4.

B

Features of Current U.S. Banknotes

This appendix describes the materials and manufacturing processes used in the production of U.S. banknotes. It discusses how these production elements combine to result in effective banknote features. The section on the vulnerability of current features describes some methods being used to simulate genuine banknote features.

CREATING UNIQUE FEATURES ON GENUINE BANKNOTES

Special papermaking and printing processes are required in the manufacture of U.S. banknotes in order to implement their design. The features of a U.S. banknote fall into three general categories: (1) substrate, (2) additional elements, and (3) the printed image. These features can be detected through visual means, through tactile means, and through a variety of detection schemes that are augmented by devices. It is important to understand how these features are produced in order to understand their interactions with one another and their effectiveness in U.S. banknotes.

Substrate

The cotton and linen paper that is the substrate for U.S. banknotes is not only a base for the printed image and additional substrate elements, it is also a feature itself. The substrate is an intentionally designed feature of currency—it contributes to the look of the note, the feel of the note, the denomination of the note, and the authenticity of the note. The distinct feel of a crisp, new bill is very recognizable.

Through engineering of the paper substrate, this characteristic feel can withstand folding, crumpling, soiling, and even laundering.

The substrate also has a distinctive look. The paper used for U.S. banknotes is a blend of two cellulosic plant fibers, flax (also known as linen) and cotton. It is supplied to the Bureau of Engraving and Printing (BEP) by a single manufacturer, Crane and Company in Dalton, Massachusetts. Processing of the substrate, which is tightly controlled, includes the use of selected plant fibers and additives and the development and use of specialized methods for pulping, washing, refining, screening, pressing, and bleaching.

The sources of raw materials are selected by optimizing quality and cost. Both the flax and cotton materials are primarily sourced from waste products from the textile industry. This approach is cost-effective because papermaking can utilize the shorter fibers that make poor thread for cloth. Currently, U.S. currency is made from nominally 25 percent flax and 75 percent cotton fibers. An assortment of papermaking chemicals—color, strength, and sizing agents—are added to the raw materials. No starch or clay agents are employed in currency paper, although these are added to most other high-quality papers to improve brightness. This difference contributes to the unique look of currency paper.[1] See Box B-1 for further details on papermaking, which convey the difficulties inherent in duplicating the "feel" of genuine currency paper.

Added Elements

A number of additional items are added to the substrate during the papermaking process. These include short fibers that are visually apparent as short red and blue threads in the paper itself. A watermark is made during the papermaking process in the $5, $10, $20, $50, and $100 notes. The watermark depicts the same historical figure as that shown on the respective bills' engraved portraits.

Higher-denomination notes also incorporate security strips, made of thin plastic embedded in the notes in the final stages of papermaking; these are marked by metallic print indicating the denomination of each note. In newer notes, the strip also contains a tiny graphic of American flags.[2] The strips have a unique position on each denomination as well as a unique fluorescent color under ultraviolet

[1]Information available at <http://www.currencyproducts.com/what_to_look_for/substrate_features.html>. Accessed March 2007.

[2]Information available at <http://www.pbs.org/wgbh/nova/moolah/anatomypaper.html> and <http://www.moneyfactory.gov/newmoney/main.cfm/currency/aboutNotes>. Accessed March 2007.

BOX B-1
Papermaking

Cellulose, the most abundant natural polymer on Earth, is the major component of all papermaking fibers, including those used in the manufacture of U.S. currency. This natural polymer is composed of long linear chains of ß-1,4-D-glucopyranose in the $^4C_1$ chain conformation with equatorially oriented hydroxyl groups as illustrated in Figure B-1-1. The degree of polymerization of these chains ranges from 15,000 for unprocessed cotton to as low as 1,000 in a bleached kraft pulp. The hydroxyl groups on these linear cellulose chains form strong hydrogen bonding networks within and between cellulose chains.

Although cellulose has four crystalline polymorphs (cellulose I, II, III, and IV) only cellulose I is found in nature. Bundles of cellulose molecules, known as microfibrils, have been shown to contain both crystalline (50 to 70 percent) and amorphous regions.

Cellulose-to-cellulose hydrogen bonds are the primary theoretical fiber-to-fiber bonding mechanism. The maximizing of the fiber surface area, fiber-to-fiber contact, and hydrogen bonding are important factors in the optimization of fiber-fiber bonding. Fiber surfaces available for bonding may be developed during the beating/refining of cellulosic pulps owing to internal and external fibrillation. In general, the greater the fiber surface area available, the greater the extent of bonding. Fiber-to-fiber contact occurs when water is removed during wet pressing and the drying process. In cellulose, hydrogen bonding occurs between hydroxyl groups.

Many different high-volume grades of virgin paper exist. In essence, the papermaking process is an aqueous-based system in which cellulosic plant fibers are first mechanically treated to remove impurities and improve subsequent process stages. The next stage is often a pulping treatment that chemically removes unwanted materials. The resulting pulp is screened, washed, and may undergo subsequent chemical bleaching. The processed cellulosic fibers may then be mechanically refined to enhance the fiber bonding capacity before introduction into the papermachine.

At the papermachine, the pulp is diluted and delivered to a porous moving belt or drum that facilitates water removal and the initial formation of a sheet of paper. In subsequent operations, presses and drying cylinders are used to remove the remaining water. In addition, papermakers utilize the papermachine to introduce an assortment of papermaking chemicals that enhance physical and optical properties of the paper. In the production of currency paper, manufacturers extend the capabilities of the papermaking process to facilitate the controlled introduction of watermarks, colored threads, security strips, and other anticounterfeiting technologies.

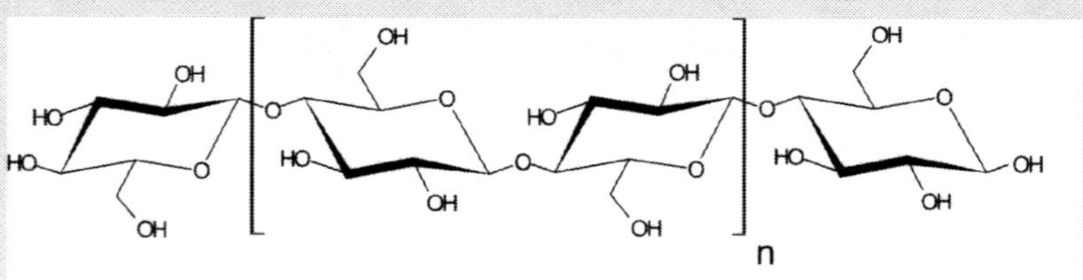

FIGURE B-1-1 The structure of cellulose.

lighting: The \$5 strip is blue; the \$10, orange; the \$20, green; the \$50, yellow; and the \$100, red.[3]

[3]Information available at <http://www.jascoinc.com/literature/pdf/appnotes/FP_01.02.pdf>. Accessed March 2007.

Specifications for the paper, including the embedded elements, ensure that U.S. banknotes have a consistent look, feel, and strength. These specifications encompass thickness, opacity, roughness, porosity, resistance to tearing, tensile strength, fluorescence, folding endurance, thread bonding, color, ash, pH, and embedded-fiber density.

Image

After the paper is delivered to the BEP as stacks of cut sheets, it is printed in a multistage printing process, front and back. Each sheet will be cut into 32 U.S. banknotes at the end of the printing process. For the $10, $20, and $50 notes, the first step is an offset process that prints a colored background simultaneously on the front and back of the sheet of notes. The background offset-printed colors are different for each denomination. The offset press is capable of maintaining register within approximately ±0.008 mm.

Next, the sheets are intaglio printed in separate steps on the front and back. The largest and most noticeable element of the intaglio-printed image is the portrait ($5, Lincoln; $10, Hamilton; $20, Jackson; $50, Grant; $100, Franklin). The portrait is also printed slightly off-center to open up space to enable the addition of the watermark and also to reduce image wear caused by folding the note in half.

On all but the $1 and $2 notes, the portrait is large enough to accommodate microprinting and fine-line details. Microprinting is used to print 0.2 mm tall letters with a line width of 0.05 mm. The production line width is approximately 0.1 mm, with a spacing of 0.1 mm. Microprinting and fine-line printing are used because they are difficult to reproduce with low-resolution electronic devices and can be viewed by the sharp-sighted or with a simple, low-power magnifier.

Inks are used in a variety of ways in banknote printing to create security features in the images. For example, infrared and magnetic patterns are incorporated that can be detected by machine.[4] Color-shifting ink is used to print the denomination in the lower right corner of the $10 notes and higher denominations. The ink is optically variable, and shifts colors on the older $10 and $100 notes from green to black, and from copper to green on the new $10, $20, and $50 notes.[5] The printed image on the note also contains "symbols of freedom" such as a torch ($10), an eagle ($20), and a national flag ($50). All of the inks used are purchased from

---

[4]Laser Technology Identifies Counterfeit Currency, in Photonics Spectra, August 2005. Available at <http://www.photonics.com/content/spectra/2005/August/applications/65824.aspx>. Accessed March 2007.

[5]More information on color-shifting inks is available at <http://www.moneyfactory.gov/newmoney/main.cfm/currency/new20#ink>. Accessed March 2007.

a single source, SICPA, headquartered in Lausanne, Switzerland. SICPA provides security inks for over 85 percent of the world's banknotes.[6]

Finally, the offset printing on the newest notes (the redesigned \$10, \$20, and \$50) contains two additional features aimed at counterfeit deterrence. Patterns and ink colors known as the banknote detection system (BDS) are used to prevent counterfeiting using color copiers. An additional digital counterfeit deterrence system (CDS) is also incorporated into the line pattern that interferes with the ability to reproduce banknotes digitally.

FEATURE EFFECTIVENESS

While it is very difficult to say with certainty which are the most important features on current notes, there are some indicators for which features are important to the different types of users of banknotes:

- The features most used by the general public are reported to be the overall "look" and the overall "feel" of the note. The discriminators reported in checking for counterfeits when a banknote looks or feels different include the watermark, security strip, color-shifting ink, fine lines, and microprinting.
- The features most used by current machine readers are the transmissive optical spectrum and printed image, magnetic patterns, ultraviolet fluorescence, ultraviolet spectrum, and the infrared properties. Low-end readers may sense only one feature; high-end readers may use 10 or more measurements to authenticate and denominate each note.
- The features most used by the blind community to authenticate notes are the tactile features. There are no features available for use by the blind public to denominate or authenticate U.S. banknotes without using a machine reader.

It is interesting to note the overlap in responses of these users; both general and blind users identified the importance of the tactile feel of the note. It is also interesting to note which features are not used. For example, most overt counterfeit-deterrent features—such as the color-shifting inks, substrate properties, watermark, security strips, microprinting, and the Federal Reserve seal—are not currently used by machine readers because they are difficult to sense, to locate, or to verify. These features are also apparently rarely used by the general public, and some are not used at all, even by experienced cash handlers.

[6] Information available from SICPA at <http://www.sicpa.com/731/764/729/752.asp>. Accessed March 2007.

On the Vulnerability of Features

A variety of vulnerabilities drive currency design. Overall, features can be vulnerable in two main modes, and the two have less to do with trends in digital equipment and more to do with criminal motivation. Table B-1 summarizes the various printing processes considered, their quality implications, spatial resolutions, and general costs. This table provides a quick review of the processes available to the counterfeiter and their relative vulnerability as well.

As has been discussed, every feature on a U.S. banknote today is vulnerable to determined counterfeiters. These counterfeiters are willing to seek out specialized materials and equipment and are willing to search the Internet for special inks, image files, and tips on which digital printers and software will make a good counterfeit. Their motivation is criminal gain, which drives the depth of their effort and the scale of their operation.

Deterring a determined counterfeiter is possible through the combination of (1) features that are difficult to simulate and (2) an educated cash-handling public that is inspired to understand the distinctive nature of these features. However,

TABLE B-1 Information Age Technologies Employed by Counterfeiters

Technology	Availability	Cost	Capability
Internet	Home	Low	Provides information, know-how, image files, access to useful materials
Thermal ink-jet printers	Home	Low	Sufficient image resolution but does not reproduce non-image features
All-in-one devices	Home	Low	Sufficient image resolution but does not reproduce non-image features
Thermal transfer printing	Home	Low	Uses smooth, glossy paper; does not reproduce non-image features
High-quality scanners	Home and office	Moderate	Captures image with sufficient resolution and no obvious artifacts
Color copiers and color laser printers	Home and office	Moderate	Sufficient image resolution but does not reproduce non-image features
High-quality digital cameras	Home and office	Low	Captures image with sufficient resolution and some image processing
Image-processing software	Home and office	Low	Easily handles the larger high-resolution files needed to counterfeit
Flatbed ink-jet printers	Commercial and printing centers	High	Sufficient resolution and use of special inks can simulate watermark, colored threads, special inks, and some level of print relief
Digital press	Commercial printing	High	Duplicates the resolution used to print currency

there are no features on currency today—issued by any country—that a dedicated counterfeiter would find impossible to simulate and pass into circulation.

Casual counterfeiters, conversely, are more easily deterred. These counterfeiters may use equipment and materials that are easily accessed to make a few "pretty good" notes on an occasional basis. They are motivated merely by the opportunity; for example, they may take unwitting advantage of the lack of banknote detection systems in all-in-one devices or cell phone cameras. Today's U.S. banknote provides only a few features that would deter such a counterfeiter. However, even if a counterfeiter makes no attempt to simulate the more difficult features, an uneducated public can facilitate their quick financial gain.

A larger threat is looming as well: the increasing availability of currency-accepting devices that remove the human cash handler from the transaction. The growth in the use of these devices is expected to eventually balance the importance of human-recognized and machine-readable features in deterring currency counterfeiting. A result of the wider availability of cash accepters is the possibility that they can more easily fall into the hands of potential counterfeiters. The would-be crooks, including hobbyists, could easily refine their methods until their notes are accepted by the reader. One possible way to address this activity is to incorporate technology in currency readers that flag such efforts.

Novel Methods to Simulate Features

Users of U.S. currency have a habitual knowledge of the color and feel of the substrate. The color of the paper is not printed but is a result of the complex papermaking process, and the feel is unique to the fibers inside and the printing on the note.

In order to incorporate genuine currency paper into their products, some counterfeiters use common household bleach on $1 notes. By masking most of the printing on the note, for example, only the denomination might be bleached away and then reprinted. This technique has been made much easier through the ability of printers to handle thicker paper and new color-matching capabilities.

The color of the traditional intaglio-printed ink and of the new offset-printed background inks is also imprinted in the subconscious of many users of U.S. banknotes. Today, the subtle colors in many currencies are created using spot color inks and pigments. Many of these are nearly impossible to reproduce with a conventional desktop printer's combination of cyan, magenta, yellow, and black (CMYK) inks. However, new software and hardware tools are emerging that can improve an RGB (red, green, blue) monitor's WYSIWYG (what you see is what you get) color capabilities. These tools are intended to enable an artist working in CMYK to more accurately repair color problems on-screen before having to surmount the hardware issues within an RGB printer's software. These tools are common among experts today.

A simpler approach to circumventing the limitations in standard color reproduction may be to empty an ink-jet cartridge and refill it with spot color ink. This could more perfectly reproduce the characteristic green that says "money" to most people. Duotone printing provides another low-cost tool to the counterfeiter.

Magnetic ink is a particularly difficult material to simulate. One solution is to incorporate nanoscale magnetic particles into suspension in today's inks. The chemistry of ink-jet inks is very complex, but it may be very reasonable to suspend magnetic nanoparticles in them for the short times—on the order of hours—needed to print hundreds or thousands of sheets.

Finally, novel methods of using image-acquisition and image-processing software can more easily simulate the features on banknotes. For example, instead of scanning an entire note and re-creating it line by line, the tools common to today's art software can enable a would-be counterfeiter to scan and edit features independently. This approach looks at selected portions of the note one at a time and is done in sizes and ways that are allowable under current usage constraints.[7] Because a counterfeiter may only scan and incorporate the features that make a note easy to recognize, many printers and copiers may not recognize the final product as a banknote. In addition, once a counterfeiter completes this task, it is a simple matter to share the image file with others via the Internet.

[7]For example, see the images on the software demonstration page at CSS Zen Garden, a demonstration of what can be accomplished visually through CSS-based design. Available at <http://www.csszengarden.com/?cssfile=/126/126.css&page=7>. Accessed March 2007.

C

Intermediate-Term Feature Descriptions

This appendix has in-depth descriptions of the innovative banknote features that could be implemented in a time frame of fewer than 7 years and that are discussed in Chapter 4 of this report. Each feature description includes subheadings dealing with various aspects of the feature:

- *Description*—An explanation of the physical principle(s) on which the feature is based. Also, the feature application as visible, machine-readable, applicable to the visually impaired, forensic applicability, and so on, is described. Furthermore, the benefits and limitations of the feature are presented; graphics may be included to depict the feature and its operation.
- *Feature Motivation*—A summary of the reasons why the feature is highly rated by the committee and reference to its uniqueness.
- *Materials and Manufacturing Technology Options*—A summary of the materials and manufacturing process that could be used to produce the feature, as well as initial thoughts on how the feature could be integrated into a Federal Reserve note.
- *Simulation Strategies*—A discussion of potential ways in which a counterfeiter could simulate or duplicate the feature and the expected degree of difficulty in attempting to do so.
- *Key Development Risks and Issues*—A discussion of the durability challenges, feature aesthetics, anticipated social acceptability, and description of the key technical challenges that must be addressed during the first phase of the development process to demonstrate the feasibility of the feature idea,

that is, to demonstrate feature capabilities and determine the usefulness of the feature in counterfeit deterrence. (The development phases are defined in Chapter 6.)

- *Phase I Development Plan*—A characterization of the current maturation level of the feature technology, key milestones to be achieved during the first development phase, and known current and planned related developments external to the Bureau of Engraving and Printing (BEP).
- *Estimate of Production Cost*—An initial assessment of additional BEP operational steps that would be required at the BEP to produce a banknote with the feature, incremental cost (higher, lower, the same) relative to the cost of the current security thread, and an indication of whether additional BEP capital equipment would be required for production.
- *References and Further Reading*—Selected references related to the feature and its associated components. Such references could include, for example, papers and conference proceedings for background on any work done relating to this feature. These lists are not exhaustive but are intended to provide a snapshot of current work related to the feature concept.

The features described in this appendix are as follows:

- Color Image Saturation
- Fiber-Infused Substrate
- Fresnel Lens for Microprinting Self-Authentication
- Grazing-Incidence Optical Patterns
- High-Complexity Spatial Patterns
- Hybrid Diffractive Optically Variable Devices
- Metameric Ink Patterns
- Microperforated Substrate
- Nanocrystal Pigments
- Nanoprint
- Refractive Microoptic Arrays
- See-Through Registration Feature
- Subwavelength Optical Devices
- Tactile Variant Substrate
- Thermoresponsive Optically Variable Devices
- Window

COLOR IMAGE SATURATION

Description

By watermarking an image, its authenticity can be assessed. In most cases, watermarking is done in one or more of the various data channels of the image. The luminance, or brightness, channel of the image has been used, as well as the frequency space of the image. Since the eye is very sensitive to luminance variations, using the luminance channel results in any image degradation or manipulation being very obvious, and watermarks—even authentic ones—are often noticeable to a human observer. The frequency channel requires considerable processing if it is used, and it can also contain image-degrading artifacts.

The technique on which this proposed currency feature—color image saturation— is based uses the saturation channel of color images to embed watermarking or other secure data. Color images are usually captured via red (R), green (G), and blue (B) data channels. Saturation is data derived from the RGB channels using various computational techniques already known in the imaging industry and does not require the development of any new technology. In watermarking via the saturation channel, no visible artifacts would generally be realized, and hence the image can be watermarked without impacting its quality in any noticeable way. The human visual system is much less sensitive to saturation channel variations (essentially color intensity) than to pure luminance variations as previously described. With the BEP capability to create, process, and print such a watermarked image, the counterfeiter would not know how the watermarking was done and, as a consequence, would be at a considerable disadvantage in attempting to create a passable note utilizing this feature.

The paper cited in the "Further Reading" section below outlines the techniques used in performing this type of watermarking. One key benefit of this approach is that it should be very robust, and it is noteworthy that only an instrument can determine the authenticity of the image so marked, since unassisted visual inspection of the note would not be adequate to authenticate it. The complexity of the authentication hardware and software is not expected to be so costly or complex that it would incur prohibitive hardware or software implementation costs.

Implementing a color-saturation feature requires that the image being used is in more than one color rather than being pure monochrome. Depending on the color model chosen for the image or on how the data are created and stored, there should be a hue, value, and chroma channel—for example, the chroma channel might be used for the watermarking. *Hue*, for example, is the color of the image, such as red, green, blue, cyan, magenta, and so on. *Value* is the brightness of the color and can be similar to luminance. The *chroma channel* is the intensity of the color, such as its "redness," for example. The eye is about 10 times less sensitive to chroma variations than to luminance variations. Color images in other encodings can be converted to have a saturation channel as required; it is just a matter of image preparation. The choice of whether the image is full color, pseudo-color, or just multicolor is optional and can be made at the time of note design. However, the image chosen would require selection based on its color characteristics and suitability for saturation channel watermarking use.

Feature Motivation

This feature would deter counterfeiting owing to the need to make an acceptable image with watermarking in the saturation channel. Also, it is expected that a watermarking scheme that would permit copying detection could be implemented. Thus, if a counterfeiter copied real currency and attempted to place the image on a counterfeit, the copied image would be detectable via the appropriate analysis mechanism. While this complete capability has yet to be verified, it would, if successful, be a very robust feature indeed. This feature is quite unique in that it uses the saturation channel of a color image to encode data; since this channel is not generally observable, a secure method of authenticating the note is provided. Furthermore, since the image is usually watermarked as a multibit-per-pixel image and then rendered as a binary image for printing via a halftoning or other binarization scheme, the would-be counterfeiter would not have access to the original image and would have great difficulty in determining from the binary image on the authentic currency the pixel values of the original image.

It is expected that this feature would deter the opportunist and the petty criminal counterfeiter and that many professional criminal counterfeiters would be highly challenged in attempting to duplicate or simulate the feature. Furthermore, the would-be counterfeiter would have to reverse-engineer the authentication hardware and software. This multitiered robustness of challenging image modification and detection methodology replication would be highly frustrating and time-consuming. Perhaps one of the strongest values of this technique is for forensic detection. The value for other users would depend on whether the technology to detect and authenticate the watermark would be shared with commercial banks or retail outlets.

Materials and Manufacturing Technology Options

A color-saturation watermark feature would be printed on the note similar to the other Federal Reserve note (FRN) features. No special processes or ink would be required. The feature's strength lies in the data encoded in the image. Since most notes do not have a full-color ink set such as cyan, magenta, yellow, and black, some effort would be necessary to develop a production process and an image that employs the inks that the Bureau of Engraving and Printing uses or is planning to use. These requirements are not restrictions but do represent design and manufacturing choices to be made.

Simulation Strategies

Duplication of this feature would require the capability of at least a criminal of professional level. The opportunist or petty criminal would be easily prohibited from making this feature work by copying. Additionally, the data encoded in the saturation watermark would be unknown to the counterfeiter, since the data would be verified by a scanning and analysis mechanism only. The design of this mechanism would not generally give away the data that it processed in order to determine authenticity.

A key characteristic of most images is that they exist in a continuous-tone or multibit-per-pixel data format. However, the printing process, such as intaglio, is a binary process in that there is either ink or no ink deposited on the substrate. There can be substantial proprietary technology in turning the continuous-tone image into the proper binary image that is capable of being rendered from a device such as an intaglio or offset printer. Again, the would-be counterfeiter would have no knowledge of the original image's continuous-tone data and hence could not readily determine either the binarization process used by the BEP or the original image data. Without this information, the would-be counterfeiter would have no idea what the original data looked like and hence could not readily determine what an acceptable forgery would look like to an analysis instrument. This kind of feature would, therefore, provide the note with substantial security advantages that are not accessible by the criminal from the currency itself. A visually similar appearance would be no guarantee that a counterfeit note would pass an authenticity examination via the scanner and processor at the point of use.

Key Development Risks and Issues

Durability

The durability of this feature should be high, since it is contained in a printed image and the image should be quite robust, as any printed feature would be. Therefore, durability is not an issue.

Aesthetics

The look and feel of the currency should not be negatively impacted by the use of this feature. The images would have to be in color or pseudo-color, and color is already present on U.S. FRNs.

Social Acceptability

There should be no issues regarding social acceptability, since the feature is just an image. However, it is conceivable that the scanning and processing of the note might cause some concern about maintaining the anonymity of currency. This feature would only facilitate the authentication of the note, however, and would not be used as a tracking feature.

Key Technical Challenges

The first key technical challenge would be to make sure that the embedding of the required data in the saturation channel did not noticeably deteriorate the image being watermarked.

The second key challenge would be to make sure that the watermark data were detectably altered if the image was copied in an unauthorized fashion. In this way, any attempt to copy and reproduce the image would cause detectable errors that would flag the currency as counterfeit.

Lastly, a scanning and processing mechanism would need to be designed that properly analyzed the note and did so at an acceptable speed coupled with tolerable cost and complexity.

Phase I Development Plan

Maturity of the Technology

This technology is modestly mature. Any required scanners and data-processing schemes are already known and tested. The only remaining issue is how well the embedded data degrade upon copying so that forgeries are easily detectable. The state of this knowledge is unknown.

Current and Planned Related Developments

No related developments in the public domain are known to the committee except as described in the reference work cited in "Further Reading," below.

Key Milestones

The key milestones required are as follows:

- Select an image or images suitable for use on currency.
- Watermark the images, scan them or copy them, and process the data.

- Investigate how well the saturation watermark passes through the currency engraving and generation process.
- In approximately 1 year following the achievement of the above results, currency could be in production, depending on resources and priorities.

Development Schedule

The committee estimates that the development of Phase I of this feature could be completed well within 2 years.

Estimate of Production Cost

Compatibility with Current BEP Equipment and Processes

The production cost should be very minimal owing to the fact that watermarking is only a printed feature. The cost impact of printing an image in more than one ink needs to be assessed, but this is likely already known.

Incremental Production Cost

The cost of this feature should be very minimal, since it is just another printed feature on the currency. For the required color image, color inks and a more complex printing process are involved, but the additional cost impact should be low to very low in the volumes of currency produced.

Required Capital Equipment

There is little in expected capital cost incurred with this feature. The need to process the watermark and scan the image would require some capital equipment, but it should be a relatively small amount. The software processing required should be capable of being developed on systems already in-house.

Further Reading

Huang, P.S., and C.-S. Chiang. 2005. Novel and robust saturation watermarking in wavelet domains for color images. Optical Engineering 44(11): 117002.

FIBER-INFUSED SUBSTRATE

Description

This proposed feature involves fiber-infused paper—that is, small-diameter optical fiber segments placed in the currency substrate. These fiber segments could be glass, acrylic, or other materials. Even metallic fibers could be placed in the substrate to radiate signals when illuminated by radio-frequency (RF) signals, for example. Optical-fiber segments, when illuminated by laser light or narrow-spectrum illumination, create a signature pattern that would be easily recognizable. This feature is envisioned as an upgrade to the current fiber content of the substrate of U.S. FRNs. To employ this feature, optical fibers, or more preferably fiber segments, are placed in the substrate. As the substrate is manufactured, these fiber segments are mixed in before the paper is dried. When the finished substrate is illuminated with light, especially laser light, the fibers light up as the incident light emanates from the ends of the fibers.

The first deterrent example would be for a user to notice the speckles of light from the substrate when it is illuminated. The mere speckles of the substrate with its embedded fibers would be somewhat complex for counterfeiters to reproduce, since the counterfeiters would have to create their own substrate. This elevates the complexity of their counterfeiting task considerably. The limitation of this approach is that anything that causes the substrate to produce visible speckles might be misconstrued as authentic. One key element of this feature is that the substrate is no longer passive when illuminated by optical or other electromagnetic radiation. The way that the substrate responds can be highly controlled.

Figure C-1 illustrates the fibers embedded in the substrate. The references in "Further Reading," below, give additional illustration of the concept and use of this technique.

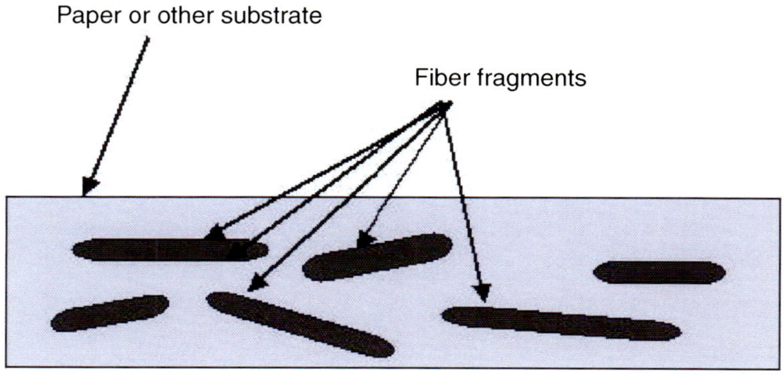

FIGURE C-1 Fiber-infused substrate.

A more robust feature would be to authenticate a note by a scan or digital photograph of the substrate and compare it with the known speckle pattern from a registered original or from data encoded on the note—this kind of longer-term feature is discussed in Appendix D.

Feature Motivation

The fiber-infused substrate feature has a good rating in the committee's analysis owing to the difficulty of implementing the cause of the feature—that is, the fibers in the substrate—and the utility of visual inspection. Furthermore, this feature would not be reproducible using electronic printing and scanning techniques and hence would frustrate a large number of would-be counterfeiters. The feature idea is also compelling owing to its requiring both the design and manufacture of the currency substrate. A counterfeiter would be challenged not only to provide a good paper substitute for authentic currency but also to build the special fibers into the substrate. This process is most likely well beyond the capabilities of all but the most dedicated and resourced operations.

This feature is quite unique, although similar techniques were used for missile verification in the Strategic Arms Limitation Treaty—SALT 1—of 1993 when fiber-embedded placards that could not be duplicated were placed on missiles. Furthermore, the costs associated with this technique would be quite low, since the cost of the materials is low, and it is their being embedded randomly that gives the technique value.

Materials and Manufacturing Technology Options

The manufacturing requirements for this feature would involve paper manufacturing and integrating the fiber fragments into the paper or other substrate material. Since the BEP's paper supplier produces the authentic substrate, it would be tasked with implementing this feature. It is not expected that this would be a difficult operation, although some tooling and process changes would no doubt be required. Once the substrate had been produced, further note production would proceed as usual.

Simulation Strategies

Simulation of this feature by would-be counterfeiters would not be easy. Furthermore, only the professional criminal or state-sponsored counterfeiter might be able to do a decent job of embedding fibers in the substrate and doing it well enough to make the operation a profitable one. Since the BEP could also control the fiber materials in the substrate, the counterfeiter would be faced with the difficult task of creating the fibers as well as making the substrate, eliminating the vast majority of criminals from attempting to do this.

Key Development Risks and Issues

Durability

The durability of this feature is unknown, but it would be dependent on the lengths of fiber embedded in the currency. If the currency was folded, the fibers could break if they were too long. Thus, the fibers would have to be short. No degradation of the fibers themselves is expected, and the only deterioration would be from breakage of the fibers if they were too long.

Aesthetics

There should be no aesthetic issues with the fiber-infused substrate feature. Unless illuminated, the note would look and feel identical to one without the feature. Even when illuminated, the feature should not detract from the note's appearance. Furthermore, the speckles that would be generated by illumination would be a comforting feature to the receiver of the note. Thus, the feature is aesthetically neutral and conforms to the look and feel of current notes, as far as is known.

Social Acceptability

There should be no issues of social acceptability surrounding this feature.

Key Technical Challenges

The key technical challenge would be the incorporation of the fiber-infusing process into the substrate production process. A key technical challenge of this feature would be the development and use of instrumentation for the analysis of the fibers. Such instrumentation could range from the simple, such as a solid-state laser diode in a penlight configuration, to the more complex, such as a small scanner that reads the currency and produces a result that could be read by a user or that gives a "go" or "no-go" signal. Such an instrument should be simple, reliable, and cost-effective, which may require a development effort, depending on what requirements are placed on the instrument itself.

Phase I Development Plan

Maturity of the Technology

The maturity level of this technology is relatively low for currency-related efforts. The science behind its use and verification is known and highly reliable, but this technique has not been implemented in high-volume, low-cost applications such as that envisioned here.

Current and Planned Related Developments

There are currently no known programs that use this feature. The papers cited below in "Further Reading" describing its use are the only ones known to relate to this effort. There may be related proprietary efforts in companies, but this is not known at present. A key issue regarding this feature is one of feature-assessment methods such as instrumentation. There may be levels of authentication methods that are desired and that are improved over time. With the increasing miniaturization of instrumentation and sensors via technologies such as microelectromechanical systems (MEMS), the state of the art is advancing rapidly and should only enhance the usability and value of this feature.

Key Milestones

The expected key milestones for Phase I would be as follows:

- Place fiber fragments in paper substrates to determine the applicability of the technique and any operational or manufacturing difficulties that might arise.
- Assess authentication techniques for a phased development of passive and active instrumentation methods over time.

Development Schedule

It is expected that achieving Phase I development would take between 2 and 3 years.

Estimate of Production Cost

Compatibility with Current BEP Equipment and Processes

Other than obtaining the substrate from the supplier, the currency-manufacturing operation should remain unchanged. The conventional intaglio printing currently used would not be impacted.

Incremental Production Cost

The capital cost of this feature on a per note basis should be quite small—likely less than $0.02—since only the cost of the fibers used would impact the substrate cost. The visual check method would have no additional cost.

Required Capital Equipment

Although there should be no capital costs for the BEP, there might be costs for the substrate manufacturer. Equipment that would prepare, add, and mix any fiber materials with the pulp would be required. The fiber material could be provided in bulk from a supplier. Post-processing of the fiber materials, such as doping and so on, could occur at the substrate manufacturer for security purposes if required. The fibers could be glass, plastic, micro, or nano materials, with custom design of the properties as required. It is unclear what the capital equipment costs would be until an acceptable fiber material design is realized. The development of the fibers would likely be coordinated with authentication technologies so that the maximum benefit from the investment is realized.

Further Reading

Chen, Y., M.K. Mihcak, and D. Kirovski. 2005. Certifying authenticity via fiber-infused paper. ACM SIGecom Exchanges 5(3): 29-37.

DeJean, G., and D. Kirovski. 2006. Certifying authenticity using RF waves. Presented at IST Mobile Summit.

National Research Council. 1993. Counterfeit Deterrent Features for the Next-Generation Currency Design, Washington, D.C.: National Academy Press, pp. 74-75 and 117-120.

FRESNEL LENS FOR MICROPRINTING SELF-AUTHENTICATION

Description

Microprinting is currently impossible to replicate using ink-jet and laser printers, but its effectiveness in deterring counterfeits is hampered by the difficulty for cash handlers—including the public, cashiers, and tellers—of verifying the authenticity of an FRN by checking for the presence and quality of microprinting. A simple loupe is probably the best way to easily distinguish counterfeit from genuine notes because of very marked differences in microprinting and fine-line details between the two. A thin lens embedded within the banknote itself can provide a simple means of self-authentication for microprinting and fine-line detail. This feature would be most useful for the general public as protection against opportunist and petty criminals. The benefit of this feature would be to make it easy to see poor-quality printing.

This feature works by cutting a hole in the paper substrate and bonding a transparent Fresnel lens over the hole. The lens is fabricated from plastic and has a curved sawtooth profile, providing the refractive index variations of a lens without the bulk (see Figure C-2). The banknote has to be bent, not folded, to position the lens about an inch above the microprinting to be viewed. Note that microprinting does not have to be text; the clock face on the $100 note is a good example of an image that can be verified with a lens. The time on the clock can be easily read using a lens on a genuine note, but ink-jet counterfeits cannot accurately reproduce the hands or the numerals on the clock. Public education would be required to alert people about what to look for and where.

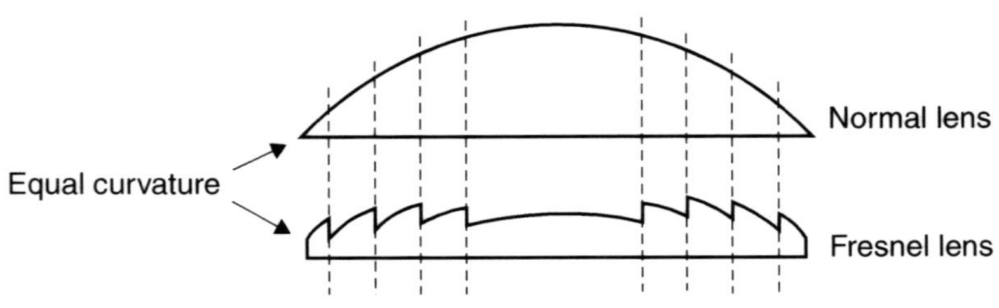

FIGURE C-2 The Fresnel lens shown here resembles a planoconvex lens that is cut into narrow rings and flattened. If the steps are narrow, the surface of each step is generally made conical and not spherical. The convex surface is reduced to concentric ridges. Fresnel lenses are flat rather than thick in the center and can be stamped out in a mold.

Feature Motivation

This Fresnel lens feature is intended for unassisted use by the general public. Cashiers and bank tellers could also use the feature, but most likely as a second line of defense because of the time required to bend the note. This feature could also assist visually impaired people in denominating their currency by using differently shaped or sized lenses for each denomination, and it could help deter "note washing" by using progressively smaller-area lenses on higher denominations. Opportunist and petty criminals should be deterred because they lack access to printing technology with sufficient spatial resolution to reproduce microprinting.

There are two risks associated with this feature: (1) home printers will be able to achieve sufficient spatial resolution to reproduce microprinting and fine-line details accurately, and (2) robust embedding of a plastic lens into the paper substrate may be difficult. The lenses themselves are robust and will work effectively even with substantial scratching and mild dirt.

This feature is expected to be cost-effective, since the cost of the mass-produced plastic lens and the cost to add a window are expected to be low.

This feature was highly rated in the committee's systems evaluation because of the likelihood that it could be easily used by the general public without requiring an external device and because it is currently difficult for opportunist and petty criminals to reproduce microprinting and fine lines. It also has potential benefit for cashiers, tellers, and the visually impaired, and could prevent banknote washing for the reuse of $5 notes as counterfeit $20s or higher. It is not expected that this feature would be of benefit to machine readers.

This feature is unique among banknotes, although a patent exists for a similar feature applied to credit cards. Plastic Fresnel lenses are commonplace as inexpensive magnifiers and are easily available in hobby stores and drugstores, although they are too thick to be embedded within current banknote substrates.

Materials and Manufacturing Technology Options

Molded plastic Fresnel lenses are commercially available and can be currently used to inspect banknotes. Fresnel lenses are commonly injected molded from PMMA plastic and are typically ~½ mm thick. Microfabrication techniques have been used to make very thin lenses (10 microns thick) but not for use in the visible spectrum.

Fresnel lenses are in common use in a variety of industries but not in banknotes. They are commercially available at craft stores and drugstores for less than $1. Holes in paper substrates have been manufactured by De La Rue, Ltd., which has also bonded metal films across these holes. This company believes that plastic can be robustly applied across paper windows. Two new processes would be required for the BEP: cutting a hole and bonding the lens to the paper. This feature enhances the effectiveness of two existing features: microprinting and fine-line printing.

Simulation Strategies

This feature can be crudely simulated by buying a cheap Fresnel lens, cutting a hole in forged paper, and gluing the lens over the hole. However, the quality of this means of reproduction is likely to be low, since available lenses are relatively thick. Microprinting cannot now be easily replicated, so even if the lens could be simulated, the magnified image would still be poor, exposing the counterfeit. If microprinting becomes easily reproducible, this feature would lose much of its effectiveness as a deterrent.

Opportunist criminals cannot simply use a computer to reproduce this feature, and therefore it may be harder for them to rationalize their actions because of the extra effort required to manually add the lens. Opportunist and petty criminals are currently limited in their ability to print with high spatial resolution, so while it might be easy for them to embed a lens, by doing so they would be adding a means for the general public to see easily the low quality of their printed note.

Professional and state-sponsored criminals should have little trouble reproducing this feature.

Key Development Risks and Issues

Durability

The key durability issue is the attachment of the lens over a hole in the substrate. Commercial banknote vendors have recently demonstrated the feasibility of attaching metallic foils over holes in paper substrates. The adhesion process should be similar for a thin plastic sheet (lens), with similar performance in durability tests. This feature looks promising as a durable feature in the short term.

Aesthetics

This feature can enhance the existing banknote by enabling users to look closely at the fine detail of the notes, but adding a plastic window may be considered unattractive by some.

Social Acceptability

No problems of social acceptability are anticipated with this feature.

Key Technical Challenges

The key technical challenge is the ability to manufacture a sufficiently thin Fresnel lens (on the order of 40 microns thick) and to attach it reliably across a hole in the banknote.

Phase I Development Plan

Maturity of the Technology

Commercially available Fresnel lenses are about 0.5 mm thick and so are unsuitable for application within banknotes. Microfabricated, thin lenses have been demonstrated in the laboratory. The committee knows of no plastic films that have yet been adhered to a paper substrate. The manufacturing readiness level is currently low.

Current and Planned Related Developments

Researchers at the University of California at Los Angeles (UCLA) have manufactured Fresnel lenses in silicon for use in the infrared using microfabrication techniques (Lin et al., 1994). Researchers at the University of Maryland have manufactured Fresnel lenses for use with x-rays using deep reactive ion etching (Morgan et al., 2004). These are not suitable for use as visual lenses as required by this feature, but they represent the current state of microfabrication. If the lenses can be manufactured from embossed plastic, microfabrication techniques may not be necessary.

Key Milestones

- Demonstrate fabrication of a Fresnel lens with appropriate dimensions. Reasonable targets are thickness equal to or less than 40 microns, aperture equal to or greater than 10 mm, and focal length about 25 mm, operating in the visible spectrum.
- Demonstrate an approach for integrating the lens over a hole in the substrate such that it passes all durability tests.
- Define a lens-fabrication process that has the potential to be affordably scaled up.

Development Schedule

The committee believes that Phase I of the development of this feature could be completed within 2 to 3 years.

Estimate of Production Cost

Compatibility with Current BEP Equipment and Processes

The committee expects that this feature would be provided by the manufacturer of the paper substrate, therefore minimally affecting BEP operations. The substrate would be delivered to the BEP with the lens already attached, and the BEP would continue to add microprinting using existing processes.

Incremental Production Cost

Microfabrication techniques should be able to provide this feature at a low incremental cost. This capability would be dependent on successful research, as outlined above.

Required Capital Equipment

The substrate provider would likely subcontract the manufacture of the lens, requiring no special equipment for the BEP or the bureau's paper suppliers. The paper supplier would need to develop a process to produce the hole.

References and Further Reading

Finkelstein, A., D.A. Dixon, and R.H. Boede. 1995. Credit Card with a Fresnel Magnifying Lens Formed in a Section of the Transparent. U.S. Patent 5,434,405. July 19, 1995.

Kingslake, R. 1992. Optics in Photography. Bellingham, Wash.: SPIE Optical Engineering Press. [Fresnel lens explanation on p. 53.]

Lin, L.Y., S.S. Lee, K.S.J. Pister, and M.C. Wu. 1994. Three-dimensional micro-Fresnel lenses fabricated by micromachining technique. Electronics Letters 30(5): 448-449.

Morgan, B., C.M. Waits, J. Krizmanic, and R. Ghodssi. 2004. Development of a deep silicon phase Fresnel lens using gray-scale lithography and deep reactive ion etching. Journal of Microelectromechanical Systems 13(1): 113-120.

GRAZING-INCIDENCE OPTICAL PATTERNS

Description

The evaluation of many optical features on currency employs the reflection of light. Most often, the light illuminating the feature impinges on the currency at near-normal incidence—that is, vertical to the plane of the substrate. The concept behind a grazing-incidence optical pattern feature is that it is intended to use light that impinges on the substrate at very high angles of incidence—that is, at about 85° to 90° from the normal—and to exploit substrate-surface irregularities, either incidental or intentional, and the patterns that they generate from the reflected light. Substrates have fairly unique characteristics at the microscopic scale and, when illuminated with either spectrally narrow or coherent light, such as from a laser, particular substrate characteristics could enable differentiation between substrates. An example of a substrate feature that could be probed this way would be an impress watermark for which the pressure from the intaglio process or other mechanical impression on the substrate used to make the watermark produced a surface relief on the substrate.

Figure C-3 illustrates the grazing-incidence model. The illumination of the substrate could be either coherent or incoherent, and the scattered light as seen by the observer could be seen as a pattern or, in the case of a watermark, as a relief image. Since substrates have unique properties that are under the control of the BEP, the scattered pattern from a banknote could be customized in a unique and secret way that would deter duplication by the counterfeiter. For instance, the intaglio process could impress highly complex depressions into the currency substrate. Additionally, the substrate surface microscopic profile could be prepressed into the substrate material so as to minimize any negative impact on the FRN production process.

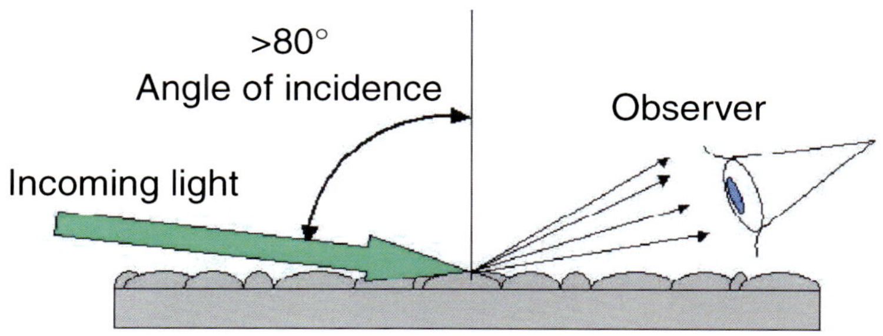

FIGURE C-3 Concept behind the grazing-incidence optical pattern feature.

Public use of this feature would require a gadget, and use by a cashier or bank teller would also be assisted.

Feature Motivation

The most compelling aspect of this proposed feature is that most counterfeiters do not produce their own substrates and therefore would be unable to begin to duplicate or simulate this feature. The currency paper supplier could properly design patterns and materials into their substrate that would make its reproduction or simulation highly complicated for all but the most determined and well-resourced counterfeiters. Different customization for different denominations would also deter the use of one FRN substrate for the counterfeiting of higher-denomination notes. This type of feature would be a deterrent for the opportunist and petty criminal counterfeiter in particular and somewhat of a deterrent for the professional criminal counterfeiter.

Another compelling aspect is that this feature is quite unique in its use of the z-axis or vertical dimension of the substrate to produce the salient effect of the feature, not ink absorption or other x, y schemas to produce the pattern or other feature characteristic.

Materials and Manufacturing Technology Options

A grazing-incidence feature is simple in that it is just an additional designed characteristic of the substrate and not produced in the printing process.

Simulation Strategies

The dedicated, very motivated, and patient counterfeiter could attempt some form of simulation of this feature, but it would require an expensive pressure plate that was engraved with a pattern that simulated the authentic feature. It is expected that only the professional criminal or state-sponsored counterfeiter would have any hope of creating even a poor facsimile of the substrate properties necessary to produce the right pattern. But it is clear that anyone wanting to re-create this feature would be greatly challenged.

Key Development Risks and Issues

Durability

This feature should have excellent durability if properly implemented. If the substrate was impressed with micro patterns over its surface, occasional crushing or folding of the currency would not damage the entire surface—making the note quite durable for general use. Only the crushing of the substrate surface, which would be hard to do completely, would abrogate the use of this feature.

Aesthetics

This feature should be aesthetically neutral, since it would not be generally observable without shining some form of illumination on it at a grazing-incidence angle on the note. Thus, normal viewing of the note would not reveal any observable pattern in general. The look and feel of the note would likely be unchanged, since the pattern or its roughness would not be noticeable to the casual user.

Social Acceptability

There should be no issues with this feature relative to privacy or environmental concerns. Properly designed, it might have characteristics that would enable visually challenged persons, since features for their benefit could also be placed on the note as part of the impression process.

Key Technical Challenges

The key technical challenges are the development of appropriate patterns and patterns that are compatible with tolerable analysis procedures. Once the patterns have been properly designed and analyzed, the production of such patterns in the currency substrate should not be difficult.

Phase I Development Plan

Maturity of the Technology

The current technology is somewhat in its operational infancy. No known implementations of this type of feature have been found in the literature.

Current and Planned Related Developments

No known development programs exist. As discussed above, most features are x, y features—that is, features such as printed patterns in the plane of the substrate. A grazing-incidence feature uses the z or vertical axis of the substrate. The committee is unaware of currency development or production programs that employ such features, other than watermarks, which use depressions in the surface of the substrate to cause images to appear when viewed vertically or nearly vertically. This feature is both illuminated and viewed at high angles of incidence.

Key Milestones

There are some key milestones to consider in the development and use of this feature:

- Design test impressions for implementation and evaluation.
- Generate impressions in substrate material using equipment that would be used for production of the substrate or note.
- Develop evaluation systems for analysis of the feature.
- Design production feature setup and testing.

Development Schedule

The first of the milestones listed above should require about a 1-year effort to design and evaluate the proper feature configuration; thus, the committee estimates that Phase I of the development of a grazing-incidence feature could be completed within 2 years, since some efforts could overlap in the development and testing process.

Estimate of Production Cost

Compatibility with Current BEP Equipment and Processes

The production cost for this feature should be quite low, since only the impression of the substrate before, during, or after printing is required. Producing this impression could be an additional step, but it is not much of a complication since the impression process is already known and operational via the intaglio printing already used.

Incremental Production Cost

The incremental production cost should be very low. Only the amortization of the wear and tear on the impression equipment incurs cost, and it is expected that this cost is already well known, since the intaglio process is well understood.

Required Capital Equipment

A small amount of optical bench equipment would be required to experiment with the use and evaluation of this feature.

Further Reading

No additional reading is suggested.

HIGH-COMPLEXITY SPATIAL PATTERNS

Description

One basic high-value attribute of current currency production is the use of intaglio presses for the printing operation. The intaglio process uses engraved or otherwise produced "masters" with which to impress the currency substrate and print the pattern. As discussed in Chapter 3 of this report, with the advent of electronic printing, scanning, and image processing, many of the high-quality features of new currency can be reasonably scanned, processed, and reproduced on ink-jet, laser-printer, and other electronic output devices. Such electronic printing, scanning, and image processing are only going to improve in the future, although the current high-quality capabilities available result in less motivation for improving these capabilities significantly in the foreseeable future. Desktop and professional publishing markets are demanding but do not require significantly higher quality than is already available. Therefore, development resources will not press current technology beyond what commercial markets require, and hence a natural performance limit has been set by market needs.

However, there is one key difference between currency production and the electronic imaging tools used by would-be counterfeiters. That difference is the analog versus digital production of the final result. Intaglio is an analog technique. One can engrave or create on the master virtually any pattern at any location on the master within the limits of the creator's art. This process is not a digital process. The electronic printer, however, is a digital process and as such has specific addressability limits. For example, a 2,500 pixel per inch printer can only lay down a pixel every 0.0004 inch. A pixel cannot be laid down 0.00027 inch from the last one because of the digitized nature of the device. Therefore, it is intended with this feature to place patterns at locations and in arrays of patterns that particularly frustrate digital printers of all kinds.

Examples of two possible patterns are shown in Figure C-4, which illustrates patterns often used for testing purposes in facsimile systems and photographic systems. While it is not necessarily suggested that these particular patterns be used, they are shown here as examples of patterns in which the gradually decreasing spacing, whether radial or lineal, will at some point cause the addressability limits of the digital printer to place two lines together without a space or two spaces without a line, and so on. With proper design and adequate intaglio/substrate printing quality, the analog system should be able to frustrate the digital system consistently.

One should notice that this feature is visible and does not require instrumentation, although a magnifier might be of some help depending on the quality of the user's visual capabilities. These features are optical in nature, and the usefulness of this high-complexity spatial pattern will depend on superior intaglio performance, line-width control, and minimum permissible line width and/or line spacing of

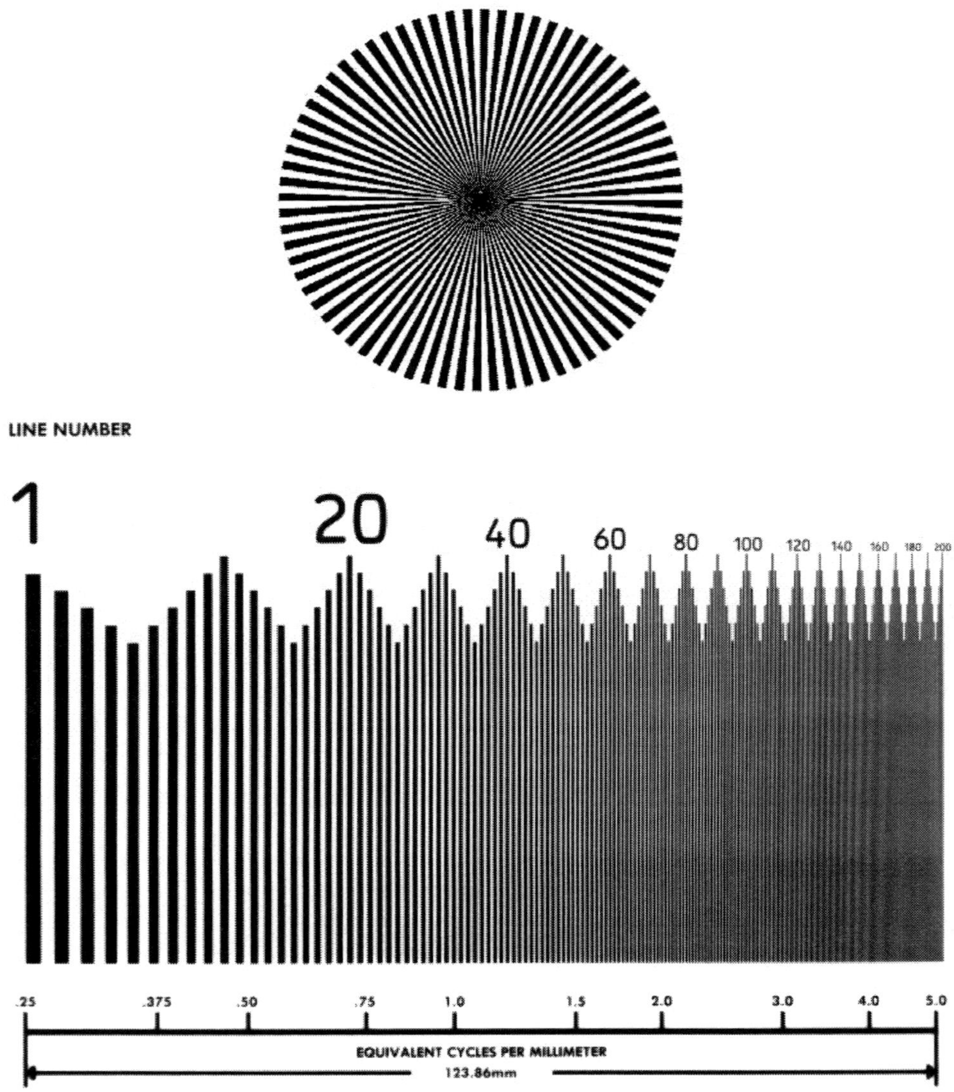

FIGURE C-4 Examples of possible high-complexity features. Upper: Starburst pattern. Lower: Sayce target.

the original currency production equipment. The basic intent with this feature is to use patterns whose full spatial bandwidth cannot be reproduced completely by digital systems and hence will show a visible defect on such systems, regardless of the image processing used. In the past, wavy patterns, chevrons, and so on could often be simulated well enough not to show errors without close examination. The type of feature discussed here is intended to create patterns in the intaglio master that cannot be reproduced in full fidelity by digital systems available on the market today or in the near future.

Feature Motivation

The high-complexity spatial patterns feature is both unique and nonunique in the following ways. The pattern is spatial and hence similar to those already included on present currency. However, it differs from current features because rather than having a fixed shape or fixed spatial frequency, this feature is composed of a multiplicity of spatial frequencies and line spacings that digital printers would be unable to reproduce all at the same time.

Currently, if counterfeiters fake a currency pattern, they might "tune" the pattern to work on their printers. Even though the resulting fake pattern does not have the same exact frequency, the ordinary user could not discern the differences without instrumentation. The inherent characteristics of this new feature prevent counterfeiters from accurately reproducing the pattern, since they cannot reproduce all the frequencies in the pattern without degrading the appearance of the pattern. Thus, the spatial band pass of the intaglio printing system exceeds that of the electronic printer owing to the latter's digitized pixel positioning requirements. The digital system is challenged by the analog nature of the original currency pattern. Clearly, a digital system could be specifically designed to have sufficient spatial bandwidth to reproduce the highly complex feature, but the current and future performance of intaglio printing and the foreseen improvements of commercial digital printer performance are such that this feature will remain a deterrent into the foreseeable future.

Should the counterfeiter decide to hand-engrave the patterns used or simulate them, he or she might succeed, with enough patience, but the effort would be detectable via pattern matching by investigators. There is the risk that digital imaging equipment will advance to the point that this feature would not have value, but if the intaglio printing process is of sufficiently high quality, as it appears to be, this is unlikely.

This feature is compelling because of several salient characteristics. First, the addition of the feature to currency uses the present intaglio printing process and hence requires little change to the present production methodology. Second, by its very nature, the pattern is designed to challenge digital scanners and printers attempting to reproduce the pattern by means other than the original intaglio technique. Third, the quality of the pattern is assessed by means of human visual capabilities and hence no equipment—just the human observer—is needed. An optical magnifier could be used for assistance, but this is a fairly simple and easily obtained tool should the user wish to have one.

The characteristics listed above make this feature a considerably attractive approach. The BEP production process would not need to be changed, and the increasing quality goals being driven by the BEP only enhance the utility of this feature. It is also a feature that does not "wear out," so to speak, and would last the life of the currency. As discussed earlier, this feature deters counterfeiting by being patterned in such a way as to frustrate digital reproduction methods. Any digital reproduction system has an inherent spatial bandwidth. When this bandwidth is exceeded, either the system cannot reproduce the signal or the signal is degraded in visually obvious ways. By creating special patterns in the analog masters for the intaglio process, it is intended to stress digital reproduction systems, which are the preferred scheme of most counterfeiters.

Materials and Manufacturing Technology Options

This feature would be an additional printed pattern on currency and hence its production would integrate well into the current manufacturing process. Few if any changes would be required to the current manufacturing process for banknotes to employ this deterrent method. Perhaps most advantageous is that as printing quality increases, the value of this feature would increase, since its complexity can be upgraded as the manufacturing quality of the currency is upgraded. Capabilities such as line-width control, line-space control, edge raggedness of printed lines, and so on all enhance the usefulness of this feature as they are improved.

Simulation Strategies

Simulating this feature would be done by using the best digital scanning and printing equipment available. However, it would be expected that the counterfeiter would be constantly frustrated by the scanning and printing system's inability to replicate the pattern in its entirety. The pattern could be designed to maximize the difficulties faced by those wishing to copy this pattern other than by re-creating it. Every imaging system has quality capabilities and noise characteristics. Building a pattern that capitalizes on the weaknesses of digital-scanning and especially digital-printing systems is key to the success of this approach. It is expected that all classes of counterfeiters would find replication of this feature difficult. Only those willing to re-create the pattern and print it using intaglio printing would have any chance at success. Therefore, state-sponsored counterfeiters might attempt to re-create the pattern, but it is expected that no one else would.

Key Development Risks and Issues

Durability

Durability should not be an issue with this feature, since it would be printed on the same currency using the same techniques now used.

Aesthetics

This feature should not degrade the aesthetics of the banknote, since the pattern can be used to print any desired image and does not need to look like a test target such as those illustrated in Figure C-4. Thus, this feature should be aesthetically neutral, would conform to the current look and feel of FRNs, and could even enhance the aesthetics of the banknotes.

Social Acceptability

There should be no issues regarding social acceptability with the use of this feature.

Key Technical Challenges

The key technical challenge with this feature is the differential quality of the intaglio spatial bandwidth and that of current and expected digital printers. If the two processes were to come to parity, this feature would have less value than if the printing process of the authentic currency continued to exceed that of digital printers. The higher the differential quality between intaglio and a counterfeiter's digital printer, the more compelling the use of this feature. Line-width control, line spacing, edge raggedness, and so on are quality metrics that would play an important role in the use of this feature. Both the jetting of ink in ink-jet printers and the fusing of toner in laser electrophotographic printers cause edge raggedness and line-width variations, weaknesses that could be exploited in the design and use of this feature.

Phase I Development Plan

Maturity of the Technology

The technology readiness of this approach is high because of the maturity of the intaglio process and its current use in FRN production. Patterns would need to be identified and tested with both the currency production equipment and the relevant electronic printers to determine the ability of the feature to perform as expected. It should be possible to perform these tests quickly and economically.

Current and Planned Related Developments

There are no known development programs for evaluating this feature, but since the process is fairly straightforward, testing its utility should be straightforward. Appropriate pattern generation could be conducted by optical test target producers already operating in the imaging industry.

Key Milestones

There are three key milestones in evaluating this feature's usefulness:

- Develop the images and evaluate the targets containing the desired spatial patterns.
- Render the desired target in a form that can be printed with the intaglio process on substrates of interest.
- Attempt to reproduce the pattern using current high-quality digital copying and/or reproduction methods.

Development Schedule

The Phase I development and evaluation of this feature should easily be carried out within a 2-year time frame. Using current high-quality digital printers and scanners would result in low development costs in the feasibility phase. The evaluation would be partly conducted using already available production equipment, so the transition from prototype to production should be fairly swift since only the creation of the intaglio masters would be required for production use.

Estimate of Production Cost

Compatibility with Current BEP Equipment and Processes

The effect on current BEP processes would be minimal, since the current process would generate the feature required. Only the costs associated with creating or engraving the original pattern would be additional. The usefulness of this feature would improve with the increased quality of output from the BEP process.

Incremental Production Cost

The cost of this feature should be very low, since it is just another printed pattern. No special inks or other equipment would be required if this feature works as envisioned.

Required Capital Equipment

For a thorough evaluation of this feature's usefulness, access to high-quality digital reproduction equipment would be required. This feature could be in one or more colors and hence color digital reproduction equipment should be available. Using the latest in reproduction equipment would allow an assessment of the robustness of this feature; it may be that this equipment is already available in-house at the BEP. If so, no special purchases would be required. Certainly, microscopes and so on would be needed to assess the feature's quality, but it is assumed that such equipment is already available to those requiring it in the BEP.

Further Reading

Johnson, J.L. 1986. Principles of Non Impact Printing. Irvine, Calif.: Palatino Press.

Smith, W. 1990. Modern Optical Engineering. New York: McGraw-Hill.

HYBRID DIFFRACTIVE OPTICALLY VARIABLE DEVICES

Description

In the most general sense, hybrid diffractive optically variable devices (h-DOVDs) appear different depending on viewing angle, and at least one part of the device operates on the basis of diffractive effects. These elements consist of diffractive optically variable devices (DOVDs)—holograms, kinegrams, exelgrams, pixelgrams, or similar structures—that also incorporate patterned reflective layers, moiré patterns, color images, nonuniform coatings, interference filters, or other features to create optical effects that are more elaborate and difficult to simulate or duplicate than those achievable with conventional DOVDs. Figure C-5 shows two examples of h-DOVDs. In sophisticated embodiments, the devices include encrypted information or hidden features that can be observed with specialized equipment, light sources or optics (for example, systems capable of retrieving phase information from a reflected or transmitted image).

The DOVD part of the device uses relief structures and/or spatial variations in the index of refraction or absorption of a material to produce diffraction patterns that create images whose appearance depends on viewing angle. The DOVD can operate in transmission or reflection modes or in both modes simultaneously. Reflection-mode surface-relief DOVDs are often coated with thin layers of metal to increase their brightness. Protective coatings are used to prevent a gradual wearing away of the relief during circulation of the note.

The other components of the h-DOVD can include printed patterns of colored

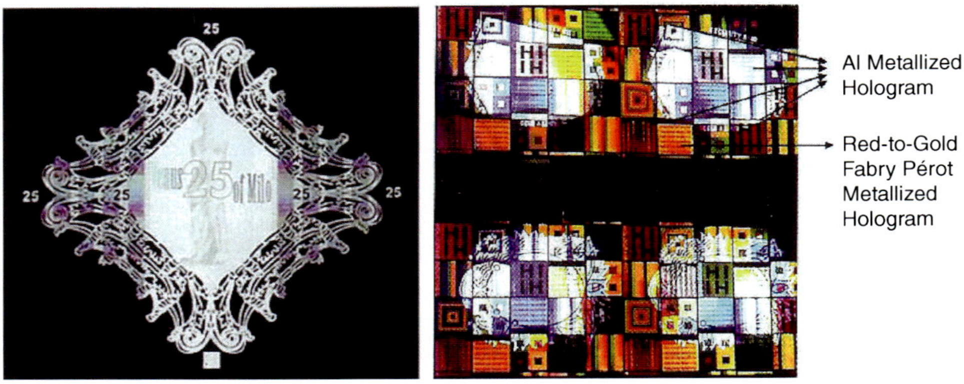

FIGURE C-5 Representative hybrid diffractive optically variable devices. Left: Thin aluminum patterns on a hologram. Right: Thin aluminum patterns and Fabry-Pérot interference filters on holograms. SOURCE: R.W. Phillips and A. Argoitia. 2005. Using roll coaters to produce anti-counterfeiting devices. Vacuum and Coating Technology 6(10):46-54.

inks, thin-film interference filters, or patterns of reflective metal formed on top of reflection-mode DOVDs or on the top or bottom of transmission-mode DOVDs. Advanced integration might involve combining color-shifting technologies such as optically variable interference-based pigments and Zero Order Devices with the DOVD.

Feature Motivation

An h-DOVD can provide both visible and hidden features to deter counterfeiting. The striking optical appearance of the device provides a distinctive visual indicator of the authenticity of a currency note. Also, the phase and other information in the images produced by the devices can be encoded and retrieved using specialized equipment that is very difficult for even sophisticated replicas or simulating structures to reproduce. All of the functionality is integrated into the currency note in a single feature, the h-DOVD, so clutter and feature proliferation are minimized.

The deterrence occurs at several levels:

- Visual inspection of a counterfeit note by the recipient can prevent its passing.
- The striking appearance and the possibility of hidden encoded information can deter attempts to generate counterfeit notes.
- The feature provides forensic information that can aid law enforcement against sophisticated counterfeiters who might be able to simulate some of the overt visual features of the devices.

Conventional DOVDs are used in more than 200 currency denominations from 78 issuing authorities worldwide, and they are widely employed in software packaging, audio and video media, tickets, and other products in which security is a concern. The implementation of more sophisticated versions that include polarization-dependent effects, encrypted images, three-dimensional volume optical effects, and other features appears to be straightforward but less well developed for commercial applications.

The nondiffractive components of h-DOVDs are also well established (for example, interference structures are used in optically variable inks, patterned reflective layers on holograms have been demonstrated, and so on), although they are typically not implemented directly with DOVDs. The combination of these features into a single device with visible and hidden functionality is less well explored. The attractiveness of this combination is that it increases the functionality and visibility of the feature while avoiding the clutter associated with separate features placed on different parts of the currency note.

Materials and Manufacturing Technology Options

The DOVDs are either created through optical exposures of photosensitive materials or through physical embossing of relief structures in thermoplastics or thermosets. Most implementations are of the surface-relief type, owing to the ease of low-cost manufacturing. Silver halide films, dichromated gelatin films, and certain classes of photopolymers are typically used for index or absorption DOVDs. In these cases, the DOVDs are written directly using an optical recording process.

For surface-relief-type DOVDs, optically or lithographically produced embossing tools can be used with many different thermoplastics or thermosets (for example, polyvinyl chloride, polyester, and so on). The metal coatings for reflective DOVDs are usually deposited by physical vapor deposition. The additional features to create h-DOVDs could consist of conventional inks (for color images, moiré patterns, and so on), vapor-deposited thin-film stacks (for interference filters), or patterns of metal deposited by physical vapor deposition (using, for example, lift-off processes with flexographically printed patterns of volatile oils). These features could be added immediately after or before the fabrication of the DOVDs. The DOVDs can be very low in cost, especially for the surface-relief type.

The additional features needed to create h-DOVDs can also be low cost, since most rely on well-established processes. The h-DOVDs can exist in the form of patches (50 percent of holograms used for currency worldwide are this type), stripes (40 percent), and threads (10 percent). These elements, like DOVDs, can be applied to the currency substrates prior to printing.

Simulation Strategies

In their conventional form, DOVDs can be simulated, at a crude level, by the use of off-the-shelf holograms (obtained from packaging materials, tickets, art supply stores, and other sources and items) that are integrated with a counterfeit by simple cutting and pasting by primitive and opportunist counterfeiters. Although not readily amenable to large-scale production, this procedure can provide, in some cases, an effective simulation strategy. The more sophisticated classes of counterfeiters can use readily available hot stamping presses and laminating equipment for the attachment of the simulated shiny strips. The hybrid nature of h-DOVDs offers an opportunity to achieve designs that provide a more unique and striking visual appearance than is possible with conventional DOVDs, thereby degrading the effectiveness of crude simulations. Also, the integration of an h-DOVD into the note substrate, for example as a woven strip, might add even more challenge to the counterfeiter.

Actual copies of the DOVD component of an h-DOVD feature can be made in several ways. For example, surface-relief-type DOVDs, which represent the lowest-cost and most widely used form of DOVDs, can be copied relatively easily by separating them from their protective coatings and then using them as tools to create copies by embossing other substrates. This method cannot be used with DOVDs that involve modulations in the index or absorption of a material. In these cases, copies can be made by using the diffracted light from a DOVD to write a new DOVD in a photosensitive material. The h-DOVD's design includes nondiffracting structures that serve to frustrate such duplication techniques and to render the simple simulation strategy ineffective.

Other countermeasures include the use of encrypted information, hidden features, polarization-dependent effects, imposed distortions, phase encoding, and other (mostly covert) features. Such attributes (both overt and covert) would make h-DOVDs difficult or impossible to reproduce accurately for all but state-sponsored organizations.

Key Development Risks and Issues

Durability

The durability of an h-DOVD is expected to be comparable with that of a DOVD or an optically variable image (OVI). Common degradation modes include wearing away of the protective coating and surface relief structures, for the case of relief-based DOVDs, and debonding from the currency substrate. Index-modulation-based diffractive elements avoid the former degradation pathway. Additional work is needed to explore issues related to durability.

Aesthetics

The emerging widespread use of DOVDs in various currencies around the world and their implementation in product packaging provide some evidence of the good aesthetic value for h-DOVDs. Suitable implementation can conform to the look and feel of U.S. currency while still providing an attention-getting feature with striking appearance.

Social Acceptability

No social concerns have been raised about the use of DOVDs in packaging, currency, or other applications. The materials used to produce these features pose no environmental concern. The same conclusions apply to h-DOVDs.

Key Technical Challenges

Technical challenges include developing optimized designs that have suitable lifetime and acceptable manufacturing cost.

Phase 1 Development Plan

Maturity of the Technology

The separate components of h-DOVDs have already been demonstrated for currency applications. Research and design are needed to define an optimized way to integrate these elements into a single currency feature that provides the desired level of overt and covert protection against counterfeiting and with acceptable manufacturing costs and aesthetics.

Current and Planned Related Developments

The production of h-DOVDs can exploit existing manufacturing capacities for DOVDs, OVIs, and other features now used in currency applications. Optimized designs for h-DOVDs are needed. Study of the primary degradation modes and lifetime of the elements is required to establish durability.

Key Milestones

There are two key milestones:

- Establish the designs and layouts and implement them in the currency substrates.
- Conduct durability tests to study the effects of wear and tear on these devices.

Development Schedule

If acceptable durability is demonstrated, this technology can complete Phase I within 2 to 3 years.

Estimate of Production Costs

Compatibility with Current BEP Equipment and Processes

The h-DOVD feature could be incorporated into the paper by the substrate manufacturer. Alternatively, some of the nondiffractive components of the feature could conceivably be overprinted by the BEP using existing processes (for example, printing for an OVI feature). The effects on the BEP operations could, therefore, be minimal, especially in the former scenario.

Overprinting with intaglio ink, for example, might require ink and/or DOVD development to ensure good adhesion to both the plastic and the cloth fibers simultaneously.

Incremental Production Cost

The cost of an h-DOVD is expected to be incrementally more than that of a conventional DOVD.

Required Capital Equipment

The capital equipment would be the same as that used for DOVDs and systems, like those for the OVI inks, that can provide the additional functionality. For patterned metallization, for example, continuous reel-to-reel systems that use flexographic printers, metal sputtering chambers, and a lift-off process could be used.

Further Reading

Chesak, C.E. 1995. Holographic counterfeit protection. Optics Communications 115: 429-436.

Colburn, W.S. 1997. Review of materials for holographic optics. Journal of Imaging Science and Technology 41(5): 443-456.

Javidi, B., and T. Nomura. 2000. Securing information by use of digital holography. Optics Letters 25(1): 28-30.

Lancaster, I.M., and A. Mitchell. 2004. The growth of optically variable features on banknotes, in Optical Security and Counterfeit Deterrence Techniques V, R.L. van Renesse (ed.), Proceedings of SPIE-IS&T Electronic Imaging, Vol. 5310, pp. 34-45.

Phillips, R.W., and A. Argoitia. 2005. Using roll coaters to produce anti-counterfeiting devices, Vacuum and Coating Technology 6(10): 46-54.

METAMERIC INK PATTERNS

Description

Patterns printed with metameric ink appear different under different illumination sources. In other words, a pattern's appearance can change depending on the color of light shining on it. In the commercial printing industry such effects are problematic and to be avoided, since having two nearby colors appear the same under one light source (for example, a fluorescent lamp) and different under another source (for example, a conventional tungsten lamp) can be quite frustrating. The currency feature proposed here would employ metameric inks to enhance any difference in appearance when the feature is illuminated with, for example, a simple light-emitting diode (LED) penlight or tungsten penlight to get patterns to show or disappear. A metameric ink feature would not always require special instrumentation for the average user, since the patterns could be designed to allow for evaluation in the home or a place of business using fluorescent or tungsten lights, both of which are commonly found.

The property on which a metameric ink feature would depend is the inks' spectral reflectances which, when combined with different illuminants, produce different colors. An extreme example of metamerism, but perhaps most illustrative of the effect, can be seen when a maroon or red car is parked in a lot that has sodium vapor lighting. Sodium vapor is used for its efficiency, since its spectral output is mainly in the yellow range, and in such illumination maroons and reds often look gray.

Figure C-6 illustrates a metameric ink pair along with a "standard" color that these inks attempt to reproduce. As described in more detail below, electronic color printers can reproduce a color that results from a metameric ink under a particular illuminant, but such printers cannot reproduce the metameric effect because the same primary colorants are used in these printers for all imaging purposes. The need to provide special inks or toners to simulate the real metameric inks would be a significant deterrent to most counterfeiters.

Feature Motivation

The idea of using inks that display metameric effects in visible light is compelling, because most counterfeiters use digital printing systems such as color ink-jet or color laser printers to generate their illicit output. However, a metameric feature is highly problematic for the counterfeiter using equipment that often uses only four printing dyes—cyan (red absorber), magenta (green absorber), yellow (blue absorber), and black. In printing a picture or other pattern and attempting to reproduce a color that is perceived as maroon, the printer combines two and

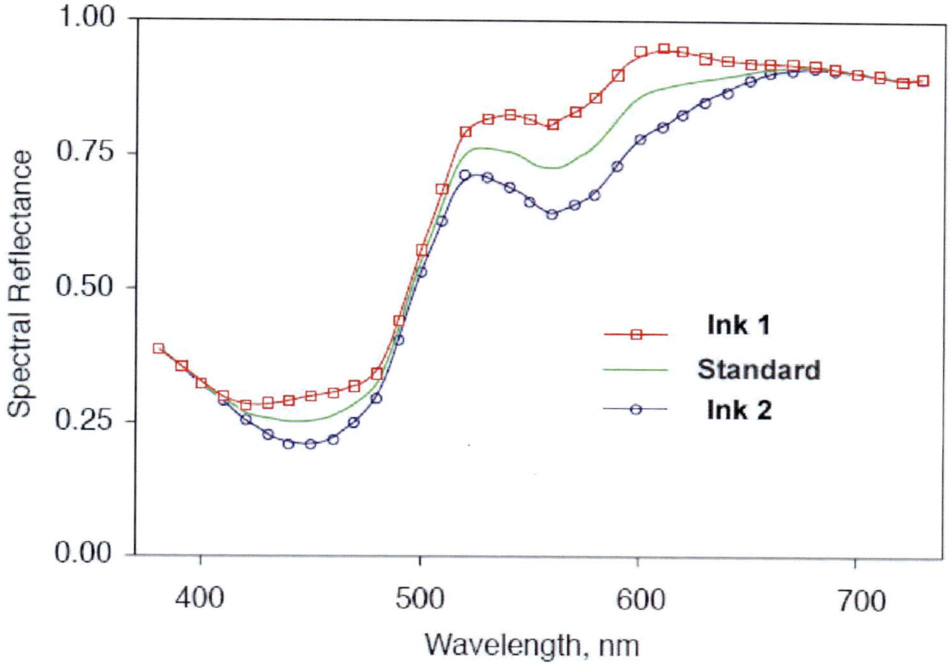

FIGURE C-6 Spectral reflectance of a metameric ink pair and a standard color ink. Notice that there are spectral differences in the 400 nanometer (nm) to 500 nm region and the 550 mm to 650 nm region. Depending on the spectral content of the illuminant—that is, its color—these inks can generate different perceived colors. In some spectral regions the lines touch or nearly touch, and in that illumination the dyes appear to be the same color. But if the inks are illuminated with 600 nm (orange) light, for example, they appear to be different colors. SOURCE: Viggiano (2004). Reprinted with permission of the Society for Imaging Science and Technology, sole copyright owners of Proceedings of the Second European Conference on Colour in Graphics, Imaging, and Vision.

perhaps three of the colored inks in some proportion to achieve the desired color. The printer might be able to produce different-looking maroon colors, but since the printing "primaries," that is, the four inks used by the printer, are the same as any other maroon, no metamerism would result. The cyan dye, for instance, is the same for all images. Unless users created their own special inks or toners in the case of electrophotographic laser printers, reproducing metameric inks would be very troublesome.

The major benefits of visible metameric features are that they use inks which already function within the BEP manufacturing process, and they could be specially designed with novel materials and chemistry to act as a significant deterrent to those who might try to duplicate the color without being able to duplicate the metameric effects. While the metameric patterns would also be useful to machine readers that could detect detailed data such as their spectra, the main advantage here is that ordinary users with little additional equipment, using either environmental illumination or inexpensive penlights, could verify the authenticity of a banknote, even in low-light conditions. The only limitation of this feature is that it would require two light sources so that the metameric effects could be seen by human observers with normal vision—either using a penlight kind of device against ambient lighting or using two common light sources in the home or workplace. It is also possible that patterns could be designed so that an observer with deficient color vision could generally detect the effects. The major color-vision deficiency in humans is red-green color blindness. Green would not need to play a role in most metameric designs.

Materials and Manufacturing Technology Options

This feature is fairly unique in that one would expect to use the latest in materials technology to create the inks for printing the metameric currency pattern(s). The use of nanocrystals and other novel chemistry for making the inks would be difficult to reproduce because a wide range of different materials would be required. While metameric inks have been used for certificates and some currency, the approach suggested here goes beyond standard metameric inks.

A compelling characteristic of this proposed feature is that including it in the manufacture of an FRN is merely a matter of printing the desired pattern with special inks using existing printing presses at the BEP. The use of special inks may or may not require an additional printing step. However, the inks would integrate well into the intaglio printing process currently used and would not require a significant change in tooling or major production processes. Laying down ink on the currency substrate is already a well-controlled process, and whether the ink is metameric or not would not alter the basic process steps in producing the currency.

Simulation Strategies

This feature would deter most counterfeiters because access to the exact inks would be problematic and simulation of this feature would require the very inks used in the currency itself. The professional criminal and state-sponsored counterfeiter would likely have access to resources for duplicating or simulating this feature; however, the other classes of counterfeiters, using digital imaging tools such as scanners, printers, and their associated software, could not print patterns that would simulate the unique characteristics of a custom-designed metameric pattern. Furthermore, trace elements in the inks or their novel nanostructure could be exploited as a forensic feature and would greatly complicate the reproduction or simulation process for most counterfeiters.

Key Development Risks and Issues

Durability

The durability of this feature should not be an issue—it is just ink on the currency substrate and would only have to tolerate ultraviolet and other types of exposure similar to what currency inks must already be able to endure. The feature would have an expected durability that does not vary much from today's ink durability.

Aesthetics

The aesthetics of the note should not be altered by the use of this feature, since the ink could be used in portraits on the currency, in chevron patterns, and so on. In fact, this feature should be aesthetically neutral and would conform to the look and feel of U.S. currency today. It is even possible that this feature might enhance the note's appearance.

Social Acceptability

This feature should be very innocuous to most users. It would be designed intentionally to look like just another printed pattern. As far as is known, there would be no privacy or environmental hazards. The actual ink chemistry might not be a foodstuff but would be designed not to emit or give off harmful effluents. Any metameric inks used would have to conform to the same safety standards that apply to current currency inks. It is not expected that this would be a debilitating issue for this feature or its proper use.

Key Technical Challenges

The key technical issues to be addressed with respect to this feature would be the design of inks that stress many if not most electronic printers to the greatest extent possible. Also, the custom-designed inks should display metameric effects under simple and available light sources, such as a portable penlight, an LED penlight, commonplace indoor lighting, and so on. Little ink would be needed per note, so the ink cost would not be expected to be an issue even if exotic materials were used in the ink chemistry. Such materials might cause the ink to fluoresce.

Phase I Development Plan

Maturity of the Technology

Little needs to be developed for this feature beyond the special metameric inks to be used—a development process that could take a few months, as the general knowledge about inks is currently at a high level. Placing special requirements on the inks could require more development time, since any such requirements are unknown at this time and they cannot be readily assessed until specified.

Current and Planned Related Developments

The committee is unaware of any related programs at the present time.

Key Milestones

There are three milestones to pass in order to properly implement this feature:

- Develop metameric inks with proper cost and durability as well as adequate metamerism in the expected user environments.
- Develop and/or identify simple light sources, such as penlights with filters or LED penlights that could be distributed and used at point-of-sale locations, with which to quickly assess the authenticity of the currency.
- Evaluate the ink's applicability to the BEP printing equipment. This effort would involve chemical interactions with the intaglio plates as well as any other chemical effects on the printing equipment.

Development Schedule

Phase I of the development of this feature could be completed within 6 to 9 months and the full development of a metameric feature is estimated to take up to 18 months, depending on equipment availability and the scale-up requirements for the ink producer. The costs for this effort are unknown but are not expected to be very high, since considerable expertise in ink chemistry is available.

Estimate of Production Costs

Compatibility with Current BEP Equipment and Processes

The compatibility of this feature with current BEP processes is expected to be very high. The efficiency of the intaglio process and the simplicity of the ink printing requirements (basically, having the substrate wet and the ink's having a higher affinity for the substrate than for the printing plate) are the basic needs for the BEP printing operation. This feature would be very adaptable to the current BEP operations.

Incremental Production Cost

It is expected that the cost would be very low because a low volume of ink would likely be used per note. There would be no unusual tools that would wear out and need to be amortized over the life of the note's production.

Required Capital Equipment

As long as the BEP is not generating the inks used in this feature, little capital cost would be required. The ink vendor might require some capital expenditures for producing the ink in sufficient quantities, but whether this would be the vendor's cost or BEP's shared cost is unknown. The development of special penlights would be placed on the penlight vendor if that identification method was chosen. There should be no BEP expense incurred in this aspect of the feature.

Further Reading

For a description of ink features used in currency production, see <http://www.currencyproducts.com/what_to_look_for/ink_features.html>. Accessed February 2007.

For a discussion of the basics of color separation by Dr. Jan Pekarovic, University of Michigan, see <www.wmich.edu/ppse/pekarovicova/071099.html>. Accessed February 2007.

Viggiano, J.A.S. 2004. Metrics for evaluating spectral matches: A quantitative comparison. Proceedings of CGIV-2004: The Second European Conference on Colour Graphics, Imaging, and Vision. Springfield, Va.: Society for Imaging Science and Technology. See <www.acolyte-color.com/papers/CGIV_2004.pdf>. Accessed February 2007.

MICROPERFORATED SUBSTRATE

Description

Transmitted light features provide security by preventing simple scanning as an image-acquisition method. Features that are not printed deliver an additional challenge to the opportunist and petty counterfeiter. Microperforation, in which very small holes are removed from the paper substrate by a laser, is one such feature.

Microperforation is a commercial technology, currently used in the manufacture of currency and the production of other secure documents. A carbon dioxide laser creates a pattern of tiny holes in the paper substrate. Microperforation is a descendant from mechanical perforation, long used to cancel checks and securities, but the use of a laser permits much smaller, more closely spaced holes. The holes are small enough that they do not significantly impact the strength of the paper, and since paper is removed rather than its being punctured, the holes do not affect the feel of the paper. They are essentially invisible in reflected light.

In currency, microperforation is typically used to form a pattern that is unique to each denomination. The pattern can be observed by means of transmitted light, either ambient light (holding the note up to a light) or, more dramatically, a light source (penlight, light table).

Because the holes are small, the viewing angle is also small. However, if the pattern is extensive enough, its existence can be verified with little effort. The details of the pattern can then be confirmed if necessary.

Microperforation is a visual feature, but it could potentially be read by a machine, either optically or conductively. It is primarily a feature for the general public but could be useful for the cash handler and teller.

Microperforation is used in a few international currencies, including the Swiss franc and some former Soviet Union currencies. Typically, the pattern of holes indicates the denomination, but a pictorial image is certainly feasible. Although not unique as a feature since it is used in other countries, microperforation is considered effective.

Feature Motivation

Microperforation rates highly for its effectiveness at deterring the opportunist and petty criminal. It is straightforward to implement in the BEP process using commercially available equipment, and costs are presumed to be reasonable. It offers advantages to both human and machine cash handlers, although as a new feature, it would require some education of human users and handlers.

While there are some simulation options with this feature, microperforation would require an additional production step in the counterfeiting process and is not amenable to simulation by printing.

Materials and Manufacturing Technology Options

Microperforation is already used in other nations' currencies, and the carbon dioxide laser is commercially available. Implementing this feature would require an additional manufacturing step at the BEP, but this step would be separate from the current process flow. Experts would need to decide at what point to add the microperforations to prevent the holes from becoming occluded during subsequent steps. Presumably, the pattern selected would reflect the denomination (perhaps even overlaid on one of the printed denominating numbers) or an image that would enhance the note design. Microperforation could be added as a design element to another transmitted light feature such as the watermark.

Simulation Strategies

Microperforation can be simulated inexpensively using mechanical punctures (a needle), but this leaves a texture that is not present in the real note and is easily detected. Microperforation can, of course, be duplicated using lasers, but this is an expensive capital undertaking, currently out of the opportunist or petty counterfeiter's range. It has been suggested that a printed image, say of small black dots, might be used to simulate a "dirty" note with holes occluded. This is not an optimal counterfeiting strategy, because, whether occluded or open, the holes in authentic notes would not be visible in reflected light. A visible pattern of dots would be a red flag to the cash handler. Further, dirt that occludes the entire pattern of holes is not typical, again alerting the cash handler.

Key Development Risks and Issues

Durability

Microperforation passes durability tests in other currency. The minimal amount of material removed in creating the holes should not impact paper strength. It is possible that holes could be occluded in dirty bills, and the pattern could become hard to see in a crumpled bill if the viewing angle changed locally owing to wrinkles. Both effects could reduce feature effectiveness as the bill became worn. Most features, however, lose some effectiveness with wear.

Aesthetics

Because the feature is not seen in reflected light, it poses little aesthetic challenge. Typical implementations use an aesthetically neutral design, such as the denomination; however, a more aesthetically pleasing design, similar to the watermark, is an option.

Social Acceptability

There are no issues of loss of privacy or environmental hazards related to the implementation of this feature.

Key Technical Challenges

The key technical challenge for microperforation is that of adding the holes at an appropriate point in the process in order to prevent hole occlusion by subsequent manufacturing steps and to permit correct placement of the feature.

Phase I Development Plan

Maturity of the Technology

It appears that this feature is technically mature, as it is in use in other currencies.

Current and Planned Related Developments

Development programs for this feature probably exist in other countries that use microperforation. Because this feature is quite simple, the development of a similar program in the United States would require relatively little investment to implement.

Key Milestones

The milestones for implementing this feature are the following:

- Design a test microperforation feature that is both aesthetically pleasing and easy to authenticate.
- Identify where to insert the microperforation step in the manufacturing process; find appropriate space and infrastructure for doing so.
- Acquire microperforation equipment sufficient for the production level required.
- Test and debug the microperforation system.
- Implement microperforation in the manufacturing process.

Development Schedule

Since this feature would rely on commercial technology already used in currency manufacture, it is likely that it could be implemented within 1 to 2 years.

Estimate of Production Cost

Compatibility with Current BEP Equipment and Processes

Microperforation technology is completely compatible with the current BEP operation, although it requires a new manufacturing step.

Incremental Production Cost

It is presumed that the incremental production cost would be low, as this feature is used in other currencies. Laser paper removal is also used in other low-cost products, such as greeting cards and cigarette papers.

Required Capital Equipment

Microperforation requires automated laser microperforation equipment sufficient for the production level required, as well as supporting infrastructure.

Further Reading

For a discussion of the application of laser microperforation in Russian rubles by Yelena Kiseleva on the Web site of the *Water Mark* magazine, see <http://www.watermark.ru/magazine/bum7.htm>. Accessed February 2007.

For a description of some commercial laser perforation systems, see <http://www.microlasertech.com/>. Accessed February 2007.

NANOCRYSTAL PIGMENTS

Description

Dispersions of semiconductor or metallic nanocrystals have the potential to provide unusual classes of inks with unique color and spectral characteristics, in both reflection and transmission modes, to form images on currency or, as embedded in a plastic matrix, to create colored features in a security strip. By synthesizing nanocrystals for specific ink formulations, complex color characteristics (measured with visible, ultraviolet, or infrared light, and in reflection, transmission, and scattering modes) can be achieved with an overall degree of control of all of these characteristics that would be difficult to duplicate using other approaches. Figure C-7 shows the kinds of colors that can be achieved using nanocrystals of CdSe, silver, and gold. An image or security strip formed with such inks could be designed to have a unique appearance as an overt feature, with unique combinations of reflecting, scattering, absorbing, and transmitting properties that would be impossible to reproduce using ordinary pigments. A key value would be the machine-readable and forensic functionality provided by the detailed spectral fingerprint, as measured in reflection, transmission, and scattering modes simultaneously, especially when implemented with features in wavelength regions (for example, the infrared) that are difficult to address with conventional organic pigments.

Metal and semiconductor nanocrystals provide optical properties (that is, absorption, scattering, and fluorescence) that depend not only on the constituent materials but on the shapes and sizes of the crystals. Bulk synthetic methods can grow such materials with well-controlled sizes and shapes, thereby providing the capability to achieve tunable optical properties throughout the visible, ultraviolet, and near infrared. Examples include gold or silver metallic nanocrystals, and semiconductor nanocrystals of InP, InAs, ZnS, ZnSe, ZnTe, CdS, CdSe, CdSe, or CdTe. These materials can be loaded into support matrices and printed as conventional inks, with the nanocrystals providing precisely defined optical properties. Nanocrystals of different sizes, shapes, or materials could be mixed to produce complex but well-defined reproducible, absorption, scattering, or fluorescent signatures. Although organic dyes might be able to simulate certain of these spectral properties, it is unlikely that they could effectively reproduce the combined wavelength-dependent scattering, absorbing, transmitting, and reflecting characteristics of suitably designed nanocrystal inks.

The feature on a banknote would consist of printed images or patterns or security strips that incorporate the nanocrystal inks. Such features would be visible to the unaided eye and would be machine-readable—the latter gathering data on the unique spectral properties of the inks. These materials might also be well suited for use in metameric inks, of the type described previously in this appendix (see the section "Metameric Ink Patterns").

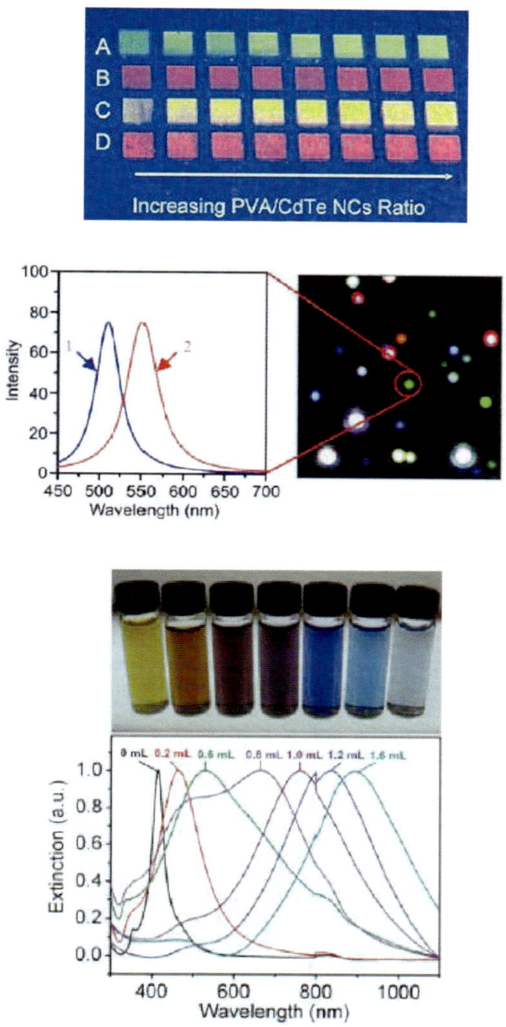

FIGURE C-7 *Top:* Photograph of ink-jet-printed squares of inks that consist of different-sized CdTe nanocrystals in polymer matrices. *Middle:* Optical micrograph of silver nanoparticles of different sizes and shapes. *Bottom:* Gold nanocrystals with different sizes and shapes and spectral extinction. SOURCES: *Top:* Tekin et al. (2006). *Middle:* Reprinted, with permission, from McFarland and Van Duyne (2003), ©2003 American Chemical Society. *Bottom:* Reproduced from Hu et al. (2006) by permission of the Royal Society of Chemistry.

Feature Motivation

As described above, dispersions of semiconductor or metallic nanocrystals provide unique color and spectral characteristics in an ink format that could be used to form images on currency with complex color characteristics (measured with visible, ultraviolet, or infrared light, and in reflection, transmission, or scattering modes) or as additives to produce color in a security strip. High levels of complexity in the appearance and spectral properties can be achieved using this approach, with a degree of control that would be difficult to duplicate using conventional dyes or with diffraction or interference-based structures. An image formed with such inks could be designed to have a unique appearance as an overt feature—such as a metameric feature as described in detail in the section "Metameric Ink Patterns," above.

A primary value of a feature printed with nanocrystal inks would be the machine-readable and forensic functionality provided by the detailed spectral fingerprint, as measured in the ultraviolet, visible, and infrared wavelength regions with a spectrometer or other optical device. These attributes could deter the passing of a counterfeit as well as attempts to generate a counterfeit.

The fact that metallic nanocrystals are already well developed and cost-effective also motivates the exploration of these materials for anticounterfeiting use, although their incorporation in inks for currency would require some development. Metallic nanocrystals, primarily gold, have been used for centuries as materials to provide coloration in items of various sorts and are especially striking because of the different colors that appear on transmission and reflection of light. Semiconductor nanocrystals are newer types of materials that need additional development for this application and have important issues such as toxicity of the materials. These crystals and the controlled means to synthesize them were discovered about 20 years ago. They are currently used for biological tagging and imaging, and some applications in security devices have been explored. To the committee's knowledge, nanocrystal inks formed from these materials have not been used in currency and therefore could be unique to U.S. currency if developed in a proprietary way.

In essence, the committee judged this feature highly because it exploits a new class of material to achieve wide-ranging optical properties in printed images or security strips. The extreme level of tunability provides the ability to define complex optical signatures that could be evaluated using suitable measurement apparatus. The high level of chemical, thermal, and physical stability in these materials also represents an advantage. Simulations of certain properties or aspects of this feature might be possible with conventional dyes and pigments, but it is unlikely that the full spectral response of the reflected, the scattered, and, possibly, the transmitted light could be reproduced without access to the specialized nanocrystal ink formulations.

Materials and Manufacturing Technology Options

Compared with organic dyes for application in currency features, semiconductor and metallic nanocrystal inks offer significant advantages in terms of brightness/contrast and stability and other attributes. For example, they do not bleach like organic dyes. In fact, the hues in centuries-old vases and goblets in which gold nanocrystals were used to generate red and green colors remain brightly visible today. The implementation in currency could involve the formulation of printable inks that consist of matrices loaded to some level with the nanocrystals. Alternatively, preformed security strips loaded with these materials, in patterned or unpatterned formats, could be integrated with the currency substrates. Certain types of metallic nanoparticles are available at low cost. The potential toxicity of these materials, particularly the compound semiconductors, must be investigated.

Simulation Strategies

Images formed using nanocrystal inks could be simulated with conventional inks, printers, and scanners to produce a counterfeit that looks, by the unaided eye and with ambient illumination, similar to an actual note. Inspection by fluorescence and/or with spectrometry equipment or in systems that evaluate the transmitted, reflected, or scattered light could, however, easily detect such simulations. Duplication would be very difficult, owing to the challenge of accurately matching the optical signatures achieved with the nanocrystal inks.

Key Development Risks and Issues

Durability

The durability of the nanocrystals (of suitable materials) is known to be good—much better than that of organic-based dyes. The durability of images formed using inks of these nanocrystals has not been explored, to the committee's knowledge. The committee expects that the durability, using properly supported matrices, could be as high as that of conventional inks that are currently used in currency applications.

Aesthetics

The nanocrystal ink feature would be aesthetically neutral.

Social Acceptability

There are serious environmental toxicity concerns associated with many of the semiconductor nanocrystals. Some of these issues are currently being addressed, as the range of applications for this class of material expands to include things such as organic LEDs for displays. Additional work on this topic is required. Funded by the National Science Foundation (NSF), the Nanoscale Science and Engineering Center at Rice University is carrying out investigations of toxicity issues with these and related classes of nanomaterials; these activities could be useful to the BEP in assessing any toxicity risk of these inks.

Key Technical Challenges

The key technical challenge is to demonstrate ink formulations that exploit the unique optical properties of the nanocrystals while at the same time enabling low-cost printing and high durability in nontoxic forms.

Phase I Development Plan

Maturity of the Technology

Gold nanocrystals have been used for centuries in various decorative and other items and therefore are a fairly mature technology, although these types of inks have not, as far as the committee is aware, been incorporated into currency. Semiconductor nanocrystals are less developed (although the synthetic procedures for forming them are nearly 20 years old) and would require more development for implementation into inks for currency applications. Toxicity issues must be investigated.

Current and Planned Related Developments

There are many industrial and academic research programs on metallic and semiconductor nanocrystals. These programs are funded by several federal agencies, including NSF, the National Institutes of Health, and the Defense Advanced Research Projects Agency (DARPA).

Key Milestones

The key milestones associated with the development of these inks are the following:

- Prepare the laboratory-scale formulation of an ink or a security strip using nanocrystals that is compatible with BEP systems.
- Provide the proof of concept of unique color and spectral characteristics.
- Overcome the toxicity issues related to certain classes of metallic and semiconductor crystals.

Development Schedule

Completing Phase I of the development of the use of metallic nanocrystal inks for currency use could be achieved within 2 to 3 years. Similar levels of development for the semiconductor inks might require twice as long.

Estimate of Production Cost

Compatibility with Current BEP Equipment and Processes

The nanocrystal ink formulations have the potential to be designed for compatibility with existing BEP printing systems.

Incremental Production Cost

The cost is estimated to be low to medium—approximately similar to that of an OVI ink.

Required Capital Equipment

Suitably designed inks could be used with existing BEP equipment.

References and Further Reading

See articles in Synthesis and Plasmonic Properties of Nanostructures. 2005. Y. Xia and N.J. Halas (eds.). MRS Bulletin 30(5).

Haes, A.J., C.L. Haynes, A.D. McFarland, G.C. Schatz, R.P. Van Duyne, and S. Zou. 2005. Plasmonic materials for surface-enhanced sensing and spectroscopy. MRS Bulletin 30(5): 368-375.

Hu, M., J. Chen, Z.-Y. Li, L. Au, G.V. Hartland, X. Li, M. Marqueze, and Y. Xia. 2006. Gold nanostructures: Engineering their plasmonic properties for biomedical applications. Chemical Society Reviews 35: 1084-1094.

McFarland, A.D., and R.P. Van Duyne. 2003. Single silver nano particles as real time optical sensors with zeptomole sensitivity. Nano Letters 3: 1057-1062.

Murray, C.B., C.R. Kagan, and M.G. Bawendi. 2000. Synthesis and characterization of monodisperse nanocrystals and close-packed nanocrystal assemblies. Annual Review of Materials Science 30: 545-610.

Tekin, E., P.J. Smith, S. Hoeppener, A.M.J. van den Berg, A.S. Sush, A.L. Rogach, J. Feldman, and U.S. Schubert. 2006. Inkjet printing of luminescent CdTe nanocrystal-polymer composites. Advanced Functional Materials 17(1): 23-28.

NANOPRINT

Description

The concept of nanoprint as a currency feature involves printed text, images, or regular arrays of patterns with critical dimensions in the micron and submicron range, and perhaps as small as tens of nanometers, formed on smooth surfaces of components of the currency, such as a windowed security strip. This feature concept represents an extension of microprint, in which the extremely high level of resolution prevents duplication by even the most sophisticated existing (or envisioned) commercial printing equipment. The technical feasibility of nanoprint derives from the discovery, through recent research in nanotechnology, of printing-like processes that offer resolution down to tens of nanometers. These approaches can be used to form nanoprint features in wide-ranging classes of materials, thereby providing an opportunity for an additional level of security. Initial implementations of these methods for nonsecurity applications indicate a potential for low-cost, high-volume operation. The nanoprint can be viewed directly using high-magnification electron microscopes—for example for covert or forensic operations. Suitable designs can produce visible collective effects for overt operations.

Feature Motivation

Nanoprint has value as a currency feature because it provides resolution and patterning characteristics that lie well outside the capabilities of the most sophisticated commercial printer systems. Owing to the small dimensions of the patterns, nanoprint can be viewed directly only by means of specialized equipment such as scanning optical microscopes. These nanoprint features can be observed indirectly, however, by exploiting visible and/or machine-readable collective effects such as diffractive or moiré effects. As an example of this mode, an array of lines with micron widths viewed through a transparent element with a similar pattern can generate moiré patterns that have spatial frequencies much lower than those of the nanoprint features themselves. By using transparent elements on the currency, such moiré effects can be generated by self-referencing with a folded piece of currency or by referencing one piece of currency to another. Such implementations would eliminate the need for a separate reading device.

In addition to feature resolution, the techniques for forming nanoprint, as described below, are compatible with a broad diversity of inks, ranging from organic molecular materials, to hybrid organic/inorganics, to nanoparticles, to biomaterials, to polymers. This materials flexibility can be exploited for additional, machine-readable forensic functionality. In particular, by combining high-resolution features with unique ink compositions, nanoprint has the potential to provide an extremely

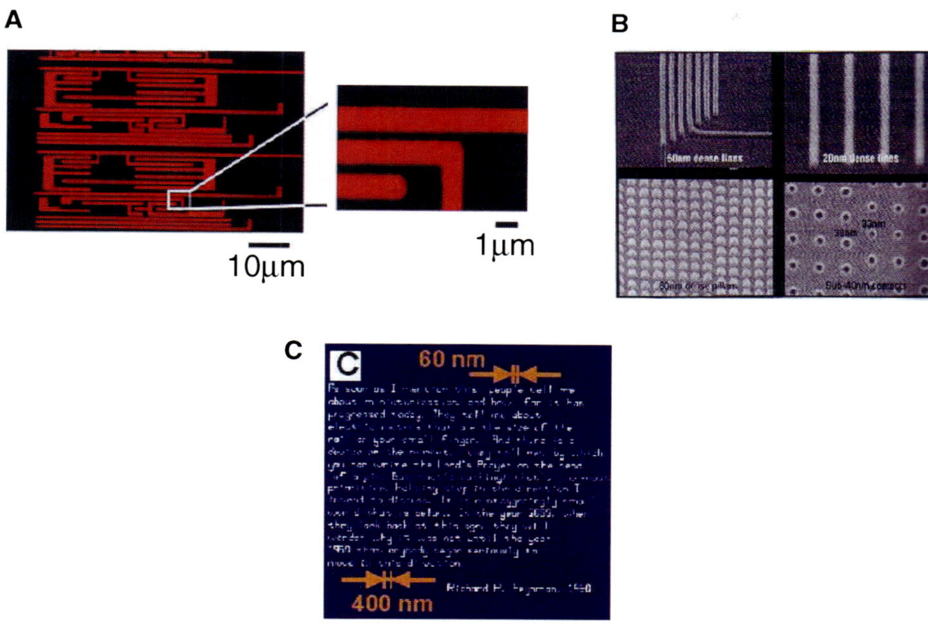

FIGURE C-8 Examples of nanoprint. Patterns of (A) fluorescently tagged proteins formed by rubber stamping, (B) polymer structures formed by imprint lithography, and (C) organic molecular monolayers formed by dip-pen writing. SOURCES: (A) Reprinted with permission from Bernard et al. (1998), ©1998 American Chemical Society. (B) Reprinted from Resnick et al. (2005), ©2005, with permission from Elsevier. (C) Mirkin (2001).

high level of security. Figure C-8 provides some examples of printing at the length scales contemplated for nanoprint, with several different classes of ink materials.

Nanoprint rated highly with the committee because it provides an extremely high level of security in a feature format that has the potential to be manufactured at low cost. The material composition of the inks and the geometries of the printed patterns can be evaluated using specialized equipment (for example, scanning electron microscopes), thereby providing valuable forensic and/or machine-readable functionality. The previously mentioned collective effects provide a strategy for the use of a nanoprint feature by the general public or by high-speed machine readers without the need for specialized microscopes.

Materials and Manufacturing Technology Options

Recent research in nanotechnology has produced several methods that can "print" structures and patterns with dimensions in the deep submicron and nanometer regime and, as described above, have the potential to create nanoprint features for currency applications. There are three different approaches that are well developed for research applications and are now in various stages of commercial development for application outside of security printing:

- *Dip-pen lithography*, in which the stylus of an atomic force microscope is used as a pen to write patterns of various materials, with lateral resolution down to tens of nanometers; this method is being developed for writing security markers on pharmaceuticals.[1]
- *Soft lithography*, in which an elastomeric stamp prints patterns of materials, with resolution in the 1 nm to 10 nm regime; this method is being commercially explored for applications in photonics and electronics.[2]
- *Imprint lithography*, in which an embossing element produces features of relief in a thermally softened polymer or a photocurable liquid prepolymer; this method is being commercialized for applications in microelectronics and optics.[3]

The manufacturing costs associated with these methods can be low, since the materials can be selected to be low in cost and the processes can be scaled for high-volume production, as evidenced by the development activities in microelectronics, photonics, pharmaceuticals, and so on. These manufacturing techniques are most effective when applied to smooth substrates. As a result, nanoprint would be most easily implemented in security strips or other structures that are separately processed and subsequently integrated into the paper.

Simulation Strategies

Nanoprint, of course, cannot be viewed directly with the unaided eye. The collective effects of nanoprint features (for example, diffraction, moiré effects) might be simulated in various ways, similar to those that are used to simulate DOVDs. The moiré effects that are possible with nanoprint are expected to be difficult to simulate accurately using structures with larger dimensions. Also, these simulation strategies would be ineffective in defeating the forensic or machine-readable functionality associated with the features themselves or the material inks.

Duplication of nanoprint would be very difficult. Although the fabrication

---

[1]NanoInk, Inc., is an emerging growth technology company specializing in nanometer-scale manufacturing and application development for the life science and semiconductor industries. See <http://www.nanoink.net>. Accessed February 2007.

[2]For example, by IBM, Inc., by Royal Philips Electronics, and by DuPont, Inc.

[3]For example, by NanoOpto Corporation, by Nanonex, Inc., and by Molecular Imprint, Inc.

techniques for nanoprint have low-cost operational characteristics, they are sufficiently sophisticated that they only exist in research laboratories and certain specialized development systems. Only state-sponsored organizations could gain access to or reproduce these systems. Even in this case, the inks themselves—for example, uniquely designed biological molecules—provide an additional level of security that could make duplication difficult even for such organizations.

Key Development Risks and Issues

Durability

The durability of the nanoprint feature is determined by the selection of the inks and could be expected to be good, especially when protective layers are used on top of the print. The committee envisions that the most natural embodiment will involve nanoprint formed on a plastic substrate that is subsequently integrated with the currency.

Aesthetics

Nanoprint, since it is not directly observable by the eye, would be aesthetically neutral. Nanoprint features that also provide collective effects (for example, moiré patterns) to generate visible features could improve the FRN's aesthetics and appeal.

Social Acceptability

There are no known concerns regarding social acceptability with respect to nanoprint.

Key Technical Challenges

The key technical challenge is to produce the nanoprint feature at acceptable cost.

Phase I Development Plan

Maturity of the Technology

The procedures based on imprint lithography currently appear to be the most well developed, with soft lithography as a close second. The approach of employing pen writing is being explored for use in security systems for pharmaceuticals.

Current and Planned Related Developments

There are many efforts in the area of nanomanufacturing at industrial and academic research laboratories. Some of these are funded by organizations such as NSF and DARPA. NSF Centers for Nanomanufacturing exist at the University of Illinois at Urbana-Champaign, UCLA/University of California, Berkeley, and Northeastern University.

Key Milestones

The key milestones for Phase I are the following:

- Develop a preliminary nanoprint feature that is visible without a device.
- Explore technical approaches to producing nanoprint features at acceptable cost on a controlled plastic substrate.
- Explore methods to integrate plastic substrate into the paper substrate with high durability.

Development Schedule

The committee estimates that Phase I of the development of nanoprint could be completed in about 2 to 3 years, with additional development time for design layouts that exploit collective effects. The key assumption is that the basic manufacturing procedures can be scaled to achieve cost-effective operation for currency applications.

Estimate of Production Cost

Compatibility with Current BEP Equipment and Processes

Nanoprint would be incorporated into the paper of the currency through security strips, plastic inserts, or related strategies. In this scenario, the nanoprint would not affect manufacturing operations at the BEP.

Incremental Production Cost

There is a potential for low cost, given the successful development of manufacturing approaches that use scaled versions of existing methods.

Required Capital Equipment

A tool for performing the printing would be needed. The systems for integrating the feature into the paper would be similar to those used in diffractive optical elements or security strips.

References and Further Reading

See articles in Materials Today. 2005. February: 50-56.

See MRS Bulletin. 2001. 26(7).

Bernard, A., E. Delamarche, H. Schmid, B. Michel, H.R. Bosshard, and H. Biebuyck. 1998. Printing patterns of proteins. Langmuir 14(9): 2225-2229.

Chou, S.Y., P.R. Krauss, W. Zhang, L.J. Guo, and L. Zhuang. 1997. Sub-10 nm imprint lithography and applications. Journal of Vacuum Science and Technology B 15(6): 2897-2904.

Michel, B., A. Bernard, A. Bietsch, E. Delamarche, M. Geissler, D. Juncker, H. Kind, J.P. Renault, H. Rothuizen, H. Schmid, P. Schmidt-Winkel, R. Stutz, and H. Wolf. 2001. Printing meets lithography: Soft approaches to high-resolution printing. IBM Journal of Research and Development 45(5): 697-719.

Mirkin, C.A. 2001. Dip-pen nanolithography: Automated fabrication of custom multicomponent, sub-100-nanometer surface architectures. MRS Bulletin 26(7): 535-538.

Piner, R.D., J. Zhu, F. Xu, S.H. Hong, and C.A. Mirkin. 1999. Dip-pen nanolithography. Science 283(5402): 661-663.

Resnick, D., S.V. Sreenivasan, and C.G. Willson. 2005. Step and flash imprint lithography. Materials Today February: 34-42.

Salaita, K., S.W. Lee, X.F. Wang, L. Huang, T.M. Dellinger, C. Liu, and C.A. Mirkin. 2005. Sub-100 nm, centimeter-scale, parallel dip-pen nanolithography. Small 1(10): 940-945.

Xia, Y., J.A. Rogers, K.E. Paul, and G.M. Whitesides. 1999. Unconventional methods for fabricating and patterning nanostructures. Chemical Reviews 99(7): 1823-1848.

REFRACTIVE MICROOPTIC ARRAYS

Description

The foundational principle of features based on the incorporation of microoptics into an FRN is that of deterring counterfeiting by presenting an easy-to-observe-and-remember visual effect, applicable to all classes of currency users, that cannot be simulated in printing alone. This feature class employs arrays of microoptic elements, lenslets, to create obvious interactive optical effects. The user typically evaluates this type of feature by moving or tilting the substrate to generate a change in the observed image (Rossi et al., 2005). The feature requires the integration of an additional material into a currency paper substrate (Bartz, 2002).

Numerous image effects are possible using microlens arrays.[4] A pair of images recorded with stereo (binocular) separation creates a three-dimensional visual perception (Iizuka, 2006; Johnson and Jacobsen, 2005). Using a sequence of images can give the appearance of action and or morphing. Completely different scenes are displayed by slightly changing the viewing direction, enabling unexpected shifts in position including reverse zooming, flipping, inverting, and distinct color or background changes. The three-dimensional perception can be combined with these tilting-induced changes. Replacing the cylindrical lenses with spherical lenses creates additional effects (Steenblik et al., 2006). In this case, tilting in either direction can cause different changes in scenes.

The detection time, with minimal or no prior training, should be less than a few seconds, and detection should be sufficiently obvious that the user is confident in the authentication. In addition to the visual aspect, the feature will have a plastic versus currency-paper tactile response and an audio response created by being scratched with a fingernail ("buzzing bee") (De Heij, 2006). Varying the location and the size of this area on the substrate of the different denominations might assist the blind in verification and reduce paper washing. The absence of surface texture should trigger further investigation of a possible counterfeit by money handlers. No human-assisted device is necessary for evaluation.

Microoptic arrays have a wide range of applications, including advertisement publishing, optical systems in photocopiers, laser printers, facsimile machines, visual displays, and high-technology fields of photonics, optical metrology, and telecommunications. With the development of low-cost plastic-production methods, advertisement publishing has become the largest-volume application. The

---

[4]Additional information is available from Human Eyes® Technology, <http://www.humaneyes.com/3d-technology/invention/>; plastic sheet supplier Spartech Plastics, <http://www.spartech.com/plastics/lenticular_sheet.html>; Epigem Limited, <http://www.epigem.co.uk/products-microlens.htm>; and SUSS MicroOptics, <http://www.suss-microoptics.com/index.html>. Accessed February 2007.

eye-catching property of these devices has led to applications on vending machine displays, magazine and CD inserts along with front covers, plastic packaging materials, greeting cards, business cards, and more. The advertisement devices use standard printing technologies directly on the back side of stiff, self-supporting plastic lenticular array stock. Printing at a resolution of only 300 dots per inch is needed for these thick-lens arrays. Desktop image analysis and publishing software is also becoming available for creating individual displays using ink-jet printed images and lens array sheets.[5] The thickness of these devices is typically 0.5 mm, which is at least 5 times thicker than the device that would have to be integrated into an FRN. Therefore, the major challenge for currency application is to achieve microoptical arrays with dynamic visual effect in a robust, at least "laundry-tolerant," plastic film integrated into the 100-μm-thick paper currency.

Feature Motivation

The microoptic array is a visual feature that goes beyond the two-dimensional printed optical image by making it interactive using three-dimensional perception and animation. A feature using microoptics can be easy to remember and use. It could also have a triggering component that is tactile and tactile/audio responsive, along with an eye-catching color design. For these reasons the committee rates these features as having high detection efficiency by the general public with no assistance and as being useful in authentication by cashiers and tellers. The tactile response of this feature might be designed to help the blind authenticate an FRN without device assistance. To implement such a feature in the thickness of an FRN requires special equipment to make microoptical three-dimensional structures and micron-resolution printing. This type of equipment is typically only available in microelectronics or similar high-technology fabrication facilities. Therefore, this type of feature will be technology blocking for all but the state-sponsored class of counterfeiter. Current state-of-the-art security document manufacturers introduced the first of this type of feature.[6] These early features have many of the properties envisioned and are compatible with high-volume production methods that will likely lead to only a moderate cost increase over the holographic stripe and embedded security strip.

---

[5]For additional information, see HP Global Solutions Catalog, <http://h30156.www3.hp.com/solutionview.cfm?SOLUTIONID=1523>, and software producer Imagiam High Image Techs, SL, <http://www.imagiam.com/content/english/lenticular_versions.htm>. Accessed February 2007.

[6]For example, the Crane and Company Motion™ currency feature uses lens arrays that give a response when tilting in both horizontal and vertical directions. The images translate 90 degrees with respect to the tilting direction. The feature is monochrome. It is based on woven thread integration with the currency. The paper and thread thickness is very smooth and uniform and thus distinct from simulations that will involve appliqués and lamination of multiple sheets. The spherical lenses in a hexagonal array weaken the fingernail scratching, "buzzing bee," audible response.

The weaknesses seen result from the use of very simple monochromatic images with significant haziness and marginal tactile contrast.

Materials and Manufacturing Technology Options

Figure C-9 shows a cross section and an oblique transverse view of an array of cylindrical lenslets, which are typical of lenticular arrays (De Heij, 2006). Each lens focuses light from specific directions to a corresponding spot on the back surface. The thickness of this device, t, for an ideal system, is typically 20 percent shorter than the focal length, f, for this system, $f = nR / (n - 1)$, where R is the radius of curvature of the refractive surface and n is the material's refractive index. Using a typical refractive index for plastic material, the focal length is about three times the radius of curvature. To obtain full areal coverage, the lenslet's pitch has to be less than $2R$. The spacing of the interlaced scenes, s, for a viewing angle change of θ is approximately $s = (t - R)\theta$. Thus, the scene height will be a fraction of the lens thickness; for a modest separation angle, θ, of 6 degrees (0.1 radian), a scene height is one-tenth the element thickness. For currency applications, the array must be less than 100 μm thick to fit physically within the thickness of an FRN. Since both focal length and the scene spacing scale with the thickness of the refractive lens, arrays will need focal lengths of 40-20 μm and the scene will need printing line widths of 4-2 μm.

The microoptic currency device must also have sufficient flexibility to pass a minimum of robustness tests such as the "laundry"-type test or similar methods proposed for examination of the wear of woven strips (De Heij, 2006). Optically

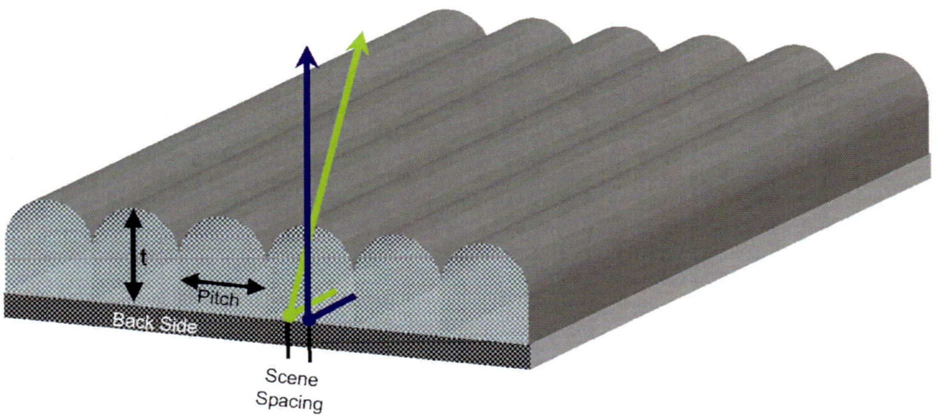

FIGURE C-9 Schematic cross section of a lenticular array showing thickness, t, pitch, and scene spacing. The blue scene is viewed straight on, while the green scene is viewed at an angle. Scenes are printed in stripes on the back side parallel to the cylindrical lenses.

transparent plastic substrates exist in currency applications in more than 23 countries. Therefore, the material robustness of a plastic material of this thickness is not likely a problem. Lenses with pitches of 50 μm are already commercially available, so embossing methods exist that can create features with nearly sufficient size; but to also achieve the comparable short focal lengths will require improvements in replication accuracy.

Although a 40-20 μm focal-length lens is just beyond the current capability of advertising-based lenticular arrays, lenslets arrays of this size are routinely manufactured for photonics device integration, fiber-optic coupling, and microelectronics (Borelli, 1999). These are made using processing methods that are slower and costly. Advances, driven primarily by consumer demand for very wide screen televisions and large-area displays, are driving down the cost of semiconductor processing lithography on large areas and prompting the development of new methods for the creation of three-dimensional structures (Chabinyc et al., 2005; Michel et al., 2002). The embossing of plastic films, which are currently used to make holographic strips at high speed and low cost, has sufficient feature line-width capability for the small pitch lenses; however, improvements in three-dimensional replication accuracy, the shape accuracy, will be needed to produce high-quality very short focal length refractive lens arrays. Eventually arrays of optical elements might be printed using advanced forms of flexography and intaglio. These printing methods already have ultraviolet and electron-beam curing[7] that would assist in transferring methods from semiconductor lithography.

The required scene-printing resolution is beyond the capability of all current ink-based color printing technologies. This resolution can be achievable using semiconductor processing lithography and related methods. A special flexographic printing process coupled with a vacuum roll coating (Phillips and Argoitia, 2005), which is similar to the security film printing methods used in currency, has the potential to achieve low-cost, high-speed production rates. Unfortunately, this method is currently only monochrome. Monochrome scenes significantly reduce the eye-catching nature of this class of feature. Even with advances in resolution, color processes using printing pigments and dyes will eventually fail because of insufficient optical absorption depth. Advances in very strong absorbers are needed. These might include microstructures such as interference filters and gratings, along with increased scene-printing resolution.

The microoptical lens arrays can be made extremely thin by the use of diffractive lenses; these are closely related to transmission holograms. Typical low-cost diffractive lens elements are made using a relief structure comprising features less than 2 μm in width and depth. Due to the binary nature of the relief and the very

[7]Electron-beam (e-beam) curing involves the curing of composite materials using a beam of highly energetic electrons at controlled doses.

flat profile, these structures are not as challenging for plastic embossing as those needed for high-quality refractive curved surfaces. Patterning volume changes in a material refractive index is another method of creating diffractive optical elements. Diffractive elements can be added to refractive lenses to reduce aberrations or used separately (Burger, 2002; Herzig, 1997). A very thin lens also would enable lens combinations that provide additional optical effects, such as a change with distance between the observer and the element and many advanced features similar to those in holographic security features. Diffractive lenses typically have lower efficiency, which leads to less-distinct and dimmer images. Further information on diffractive elements is presented in the section above, "Hybrid Diffractive Optically Variable Devices."

In addition to the refractive and diffractive microoptic lenses described above, microstructures with size and shape on the order of the wavelength of light, 0.5 μm, can create numerous additional optical effects. Examples of these effects that do not have printed two-dimensional equivalents include spectral dispersion (splitting of white light into a rainbow of colors) using gratings; angle-dependent high reflectivity or opaqueness using total internal reflection; advanced optically variable color shifts using interference filters; image offsets using Fresnel prisms or other wave-guiding structures; glistening (like a cut diamond) by creating a microstructure in a very high index material; and opalescence, with artificial photonic crystals.

The micron printed optical material needs to be integrated with the paper substrate. Most current currency with holographic plastic strips uses hot pressed stamping to attach plastic strips on the surface, as shown in Figure C-10(A). These are very thin plastic films. The thickness of refractive microoptical elements makes this method impractical. A newer approach, shown in Figure C-10(B), weaves the strip into the paper substrate, creating regions in which the plastic strip is buried and other regions in which it is exposed to the surface. In the woven-thread approach, a 20 μm to 40 μm thick plastic strip will compromise paper robustness, especially near the edges of the plastic windows. The device, which only needs to be thick in the windowed areas, suggests an approach similar to that shown in Figure C-10(C) to improve robustness.

The intaglio printing creates 5 μm to 10 μm thick ink that can be formulated to adhere to plastic and paper. Using this ink to print across the boundary might reduce delaminating. Intaglio printing over the microoptic element could also be used to fill in selected areas of the microstructures, which will nullify the optical effect in the inked regions. Besides this additive three-dimensional structure, the high-pressure intaglio printing combined with deformable plastics could create additional patterned optical functions. These novel enhancements to the microoptical base feature would benefit from higher-resolution intaglio printing that might be-

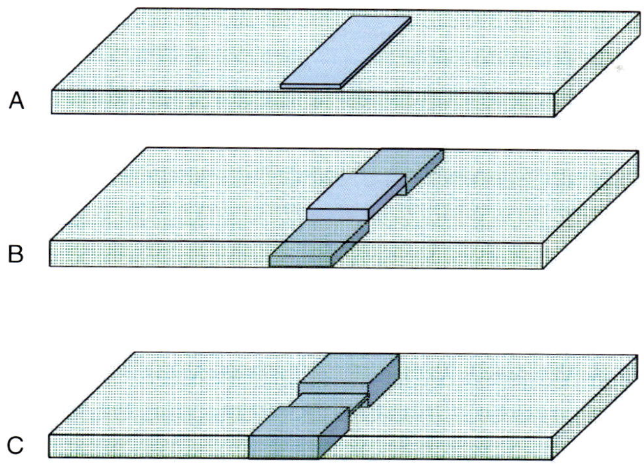

FIGURE C-10 Schematics of methods of integrating plastic material (blue) into a paper substrate (green). (A) A thin plastic film attached to the surface. (B) A woven security strip used for thicker films. Regions of the film are buried within the paper, and others regions are exposed to the surface. (C) A variable-thickness windowed strip in which the device regions are thicker, while the other regions are buried in the paper. The thin regions could be created by laminating the optical devices onto a plastic security strip such as that already used in U.S. currency. This approach also has the advantage of easily variable shaped windows.

come possible with changes in curing methods[8] combined with higher-resolution plates made using laser engraving (Deinhammer et al., 2006).

The shape of the variable-thickness area is easily controlled, as exemplified in Figure C-10(c), and its contrasting surface texture, owing to the material change in the windowed area, creates a tactile feature that is easily made note-specific. This type of feature might be effective for the blind. It also acts as an enhanced trigger for all currency users and can be used discretely if needed.

The microoptical array also has an audio signature when scratched with a sufficiently sharp fingernail, pen, or pencil. This feature is similar to the scratch feature on the euro (De Heij, 2006). The audio effect is stronger for cylindrical lens arrays than for spherical. A fingernail is sufficient for even 40 μm pitched arrays.

Simulation Strategies

Printing methods are not expected to be able to simulate the optically variable display, the three-dimensional perception, the audio signature, and the tactile feel

[8]Sun Chemical WetFlex™ technology with Energy Sciences, Inc., <http://www.ebeam.com/>, and <http://www.specialchem4coatings.com/news-trends/displaynews.aspx?id=5368>. Accessed February 2007.

of this feature. The latter two characteristics of the feature could be simulated by add-on artwork, but since these would not be integrated into the substrate, they should be obvious if tested.

It is expected that although the state-sponsored counterfeiter would find the optically variable feature difficult to produce, without the cost pressure of high production rates this feature would not likely deter production by this class of counterfeiter. The state-sponsored class of counterfeiter also has the ability to integrate the optical device with the substrate.

The counterfeits of lower quality, typical of those made by the professional criminal down to the opportunist, would not match the tactile response accurately and would completely miss the optical effects. For example, a hot stamped holographic strip might yield an optically variable feature but would lack the physical integration of the optical material with the paper. Also, the holographic simulation would not have the audio signature. Printed image simulations would be static. The lower class of simulation might use an advertising-type microoptic array, which would be so thick that it would look and feel as if it was glued onto the substrate.

Simulating the machine authentication at the level of near-infrared signature and simple optical signature might be possible for the professional criminal counterfeiter class. Either simulation would require proprietary knowledge of the tests methods designed into the machine or very extensive reverse engineering or experimentation.

Key Development Risks and Issues

Durability

The plastic used and the adhesion of this material to the paper have been worked on during the creation of the embedded security strip. A thin plastic strip has been demonstrated; however, this device needs to be as thick as practicable to enhance the working of the device. Likely failure will be on the paper edges of the interface with the plastic.

To counter the bleaching of intaglio ink on an FRN, a plastic that has chemical behavior similar to that of the cured ink would be advantageous. In addition, if the ink from the intaglio printing was used as an adhesive bond between the plastic and the paper, once that ink was removed the likely resulting separation of the feature from the substrate would render the substrate useless for the counterfeiting of higher denominations.

Aesthetics

The feature has the capability of engaging the user's attention without making the image complex and intricate. This type of feature has the potential to enhance the aesthetic appeal of U.S. currency.

Social Acceptability

This feature is a human-interfaced device and should not present any concerns over loss of privacy. The materials used are already in current U.S. currency. However, switching to laser-engraved intaglio plates should reduce the use and cost of disposing of water-soluble nickel salts. If it can be made compatible with FRN substrate materials, switching to e-beam cured inks might enable a reduction in the emission of volatile organic compounds. Ultraviolet acrylic ink would likely add literally more headache at the BEP.

Key Technical Challenges

An advanced feature in this class should be close to that used by advertisers for catching and holding the consumer's attention. The visual effect should have an eye-catching three-dimensional perception effect, and the need to tilt the note to see the feature's effects should promote interaction with the user. The use of color shifting would help make the feature harder to simulate by requiring even higher-resolution printing. The fact that two different materials are integrated into the note should be used to enhance the tactile contrast. The use of structures that enhance the nail-scratching buzzing bee sound and tactile response might help the blind.

Commercially available plastic microoptic arrays that yield bright sets of images are typically four times thicker than an FRN substrate.[9] Several technical challenges exist in reducing the thickness without significantly increasing cost or compromising functionality. Similar challenges exist in advertisements and security features for advance packaging, consumer product labeling, a potential desktop-publishing three-dimensional photography product, and fabrication of large-area displays (Levinson, 2005). Arrays with the basic functionality needed for currency application are easily made in sophisticated microfabrication facilities with slow and expensive processes (Borrelli, 1999). Printing of the image on the back side of

[9]Additional information is available from Human Eyes® Technology, <http://www.humaneyes.com/3d-technology/invention/>; plastic sheet supplier Spartech Plastics, <http://www.spartech.com/plastics/lenticular_sheet.html>; Epigem Limited, <http://www.epigem.co.uk/products-microlens.htm>; and SUSS MicroOptics, <http://www.suss-microoptics.com/index.html>. Accessed February 2007.

these arrays would require line widths of only a few microns. This is not currently possible with high-speed commercial printing processes. Expensive lithographic methods similar to those used in microelectronics fabrication achieve this resolution, however with limited color selection (Phillips and Argoitia, 2005). Additional optical effects besides dye and pigment-based coloring might be needed to achieve a full-color display in currency.

These are incremental improvements beyond what is becoming available from currency substrate suppliers. The key challenges will be durability and cost. The durability issues will likely limit the use of this type of feature to high-denomination notes. Improvements in the integration of paper and plastic films will be important. Challenges include efforts to do the following:

- Add a color three-dimensional perception, which will require printing at micron line widths or integrating advanced microstructure optical devices.
- Achieve robust integration of the microoptical device with the paper substrate, enabling varying size, shape, and positioning of the windows.
- Increase the fingernail scratching audio and tactile effects.

Commercially available lenticular plastic elements can be obtained as single sheets and continuous web rolls produced by extrusion, embossing, molding, and casting. These are available in a variety of optical-quality plastic materials with lens pitches from 1.7 mm to 40 μm. A design challenge for these devices is to minimize aberrations, ghost images, and out-of-focus scattered light. A thinner, more-flexible device would enable additional applications in packaging advertisement and might create a three-dimensional computer printing photography market creating unaided three-dimensional displays made on the home office computer printer. Individual consumers could record images using digital cameras with just the addition of a special lens. The images recorded could then be printed on special lenticular stock, viewed with animation effects on ordinary computer screens, sent in e-mails, and possibly viewed statically on specially modified computer displays. No special glasses would be needed for viewing just a "special paper" for ink-jet or thermal transfer printers. This is not much different from current practice.

The typical array with 16 images resolved in a ±30 degree tilt using a 525 μm thick lens array will need interlaced printed scenes of 50 μm. This is in the range of current printing resolutions, including flexographic printing on thin plastics and desktop-publishing printers. Thinner devices with corresponding shorter focal lengths need proportionally narrower scene lines, which will challenge existing high-speed and desktop-publishing printer resolution. Paper alignment in the printer will need improvement in order to print reliably directly on these thin arrays. Current lens array stock for individual consumers is a few dollars, but advertisement material cost is approaching that of high-quality photo paper.

The advertisement elements are typically freestanding, made with stiff materials. Flexibility will be the consequence of thinner elements, and it would be further improved by using different materials. The thinner elements with the shorter focal lengths will require higher replication accuracy to maintain the shape of the lens surface.

Phase I Development Plan

Maturity of the Technology

The first introduction of an example of this feature-class product has occurred. Prototypes are available for testing, and medium-scale manufacturing has started. Future improvements are under development. The small-scale manufacturing will yield real-life product experience.

Current and Planned Related Developments

Knowledge gained from the first product introduction should be used to refine and improve the design for the FRN.

Key Milestones

- Define and show the initial feasibility of printing optical elements at low cost. This effort would include demonstrating a high-resolution stamping process for optical elements of this size or embossed films. Determine if sufficient resolution exists to print simultaneously on the reverse side while also attaining the required registry, and explore a viable approach for printing the elements on a high-speed drum printing flexographic/offset press.
- Explore approaches to adding color three-dimensional perception to the optical effect. This would include a pigment and dye approach, or, alternately, an interference filter stack; high printing resolution; and the integration of advanced dye or pigment with high-resolution offset printing.
- Develop robust integration of a microoptic device with the paper substrate, including laminated plastic film optical elements to achieve the required durability.
- Explore creating varying window sizes, shapes, and positioning.
- Investigate increasing the tactile effect.

Development Schedule

The committee believes that Phase I of the development of this feature could be completed within 2 to 3 years.

Estimate of Production Cost

Compatibility with Current BEP Equipment and Processes

This feature would have to be outsourced to the substrate manufacturer. Minimal changes in BEP operation would be needed. The BEP would play an important role in refining the robustness requirements and product improvements. New testing and quality-control measures would be needed.

Incremental Production Cost

From the BEP perspective, this material should have a mature product cost 1.5 to 2 times that of holographic strips and significantly higher than that of the current security strip. The holographic strips in high-volume production cost $\$7/m^2$ to $\$9/m^2$ in 2004, which amounts to $6 to $7 per 1,000 FRNs. Therefore, this $9 to $12 per 1,000 note feature is classified as having a high incremental cost.

The feature's development program and the resulting high-volume cost schedule should follow a trail similar to that of the development of the euro holographic strip, which dropped in cost from $12 per 1,000 notes to $6 to $7 per 1,000 notes in 5 years. For a production rate of 1 billion notes per year, the cost amounts to an estimated $30 million for development. The capital equipment cost for a rate of 1 billion notes per year with an 8-year fixed depreciation and assuming a 100 percent margin above incremental cost suggests that a capital investment of between $20 million and $30 million would be required. Very high resolution vacuum roll coating demetallization equipment costs about $20 million installed. The embossing and plastic laminating equipment are likely less expensive but still approach $10 million installed each. Therefore, an initial cost for this type of feature will be in the range of $12 to $16 per 1,000 FRNs, with an outlook of $9 per 1,000 notes.

Printing equipment is likely about the same capital investment but with production speeds 10 times higher. Replacing the optical element formation and or the micron line process with a printing process would have significant impact on cost reduction.

Required Capital Equipment

The BEP would be required to acquire additional testing and quality-control equipment, and if advanced intaglio was required, major capital for press replacements would be needed. If printed optical elements are needed, the film will likely be printed after integration into the substrate, and this would require the new printing process equipment.

References and Further Reading

Bartz, W.J. 2002. Durability of optically variable devices on bank notes. Proc. SPIE, Optical Security and Counterfeit Deterrence Techniques IV, R.L. van Renesse (ed.), Vol. 4677, pp. 81-88.

Borrelli, N.F. 1999. Microoptics Technology: Fabrication and Applications of Lens Arrays and Devices. New York: Marcel Dekker.

Burger, R.J. 2002. Lenslet Array Systems and Methods. U.S. Patent 6381072, April 30, 2002.

Chabinyc, M.L., W.S. Wong, A.C. Arias, et al. 2005. Printing methods and materials for large-area electronic devices. Proc. IEEE 93(8): 1491-1499.

De Heij, H.A.M. 2006. Public feedback for better banknote design. Proc. SPIE, Optical Security and Counterfeit Deterrence Techniques VI, R.L. van Renesse (ed.), Vol. 6075, pp. 1-40.

Deinhammer, H., D. Schwarzbach, R. Kefeder, and P. Fajmann. 2006. The implications of direct laser engraved intaglio plates on banknote security. Proc. SPIE, Optical Security and Counterfeit Deterrence Techniques VI, R.L. van Renesse (ed.), Vol. 6075.

Herzig, H.P. 1997. Microoptics Elements, Systems and Applications. Philadelphia: Taylor and Francis, Ltd.

Iizuka, K. 2006. Welcome to the wonderful world of 3D. Optics and Photonics News 17: 43-51.

Jang, S.-J., S.-C. Kim, J.-S. Koo, J.-I. Park, and E.-S. Kim. 2004. 100-inch 3D real-image rear-projection display system based on Fresnel lens. Proc. SPIE, Integrated Optical Devices, Nanostructures, and Displays, Keith L. Lewis (ed.), Vol. 5618, pp. 204-211.

Johnson, R.B., and G.A. Jacobsen. 2005. Advances in lenticular lens arrays for visual display. Proc. SPIE, Current Developments in Lens Design and Optical Engineering VI, P.Z. Mouroulis, W.J. Smith, R.B. Johnson (eds.), Vol. 5874, pp. 59-69.

Lancaster, I.M., and A. Mitchell. 2004. The growth of optically variable features on banknotes. Proc. SPIE-IS&T Electronic Imaging, Optical Security and Counterfeit Deterrence Techniques V, R.L. van Renesse (ed.), Vol. 5310, pp. 34-45.

Levinson, H.J. 2005. Principles of Lithography. 2nd ed. SPIE Press Monograph Vol. PM146. Washington, D.C.: SPIE—The International Society for Optical Engineering.

Michel, B., A. Bernard, A. Bietsch, E. Delamarche, M. Geissler, D. Juncker, H. Kind, J.P. Renault, H. Rothuizen, H. Schmid, P. Schmidt-Winkel, R. Stuts, and H. Wolf. 2002. Printing meets lithography: Soft approaches to high-resolution patterning. CHIMIA 56(10): 527-542.

Phillips, R., and A. Argoitia. 2005. Using vacuum roll coaters to produce anti-counterfeiting devices. Vacuum Technology and Coating 6(10): 46-53.

Rossi, L.M., J. Saarinen, and M. Gale. 2005. MICRO-OPTICS: Micro Optics technology fulfills its promise. Laser Focus World. October; <http://lfw.pennnet.com/articles/article_display.cfm?article_id=238630>. Accessed March 2007.

Steenblik, R., M.J. Hunt, and M.E. Knotts. Nanoventions, Inc. 2006. Micro-optics for Article IdentificationCrane, U.S. Patent 7006294, February 28, 2006; <http://www.nanoventions.com/nano4/index.html>. Accessed March 2007.

Watson, B. 1999. Science makes a better lighthouse lens. Smithsonian 30(5): 30. Available at <http://www.smithsonianmag.com/issues/1999/august/object_aug99.php>. Accessed March 2007.

SEE-THROUGH REGISTRATION FEATURE

Description

For a see-through registration feature, one side of a substrate aligns with an associated feature on the reverse side—that is, when the FRN is held up to a light source, the combination of the two features on the two sides of the note gives rise to a new "complete" feature.

With a non-opaque substrate held up to a light source, features on both sides of the currency can become connected in a significant and obvious manner, becoming a new feature. Types of registration can include but are not limited to the following: (1) the continuation of a line shape; (2) two (individually nonrecognizable) parts of an image or design that together render a complete image; and (3) lines or shapes that line up exactly with identical features on both sides of the FRN. These features can also be combined to create a transmitted color different from either of the printed-image colors. The acuity of human vision has an impressive and rapid ability to discern between good and bad registration—about 10 seconds of arc, meaning ~0.03 mm from 60 cm.

In order for this feature to be most effective, the images involved must be large enough to attract the user's attention, and the design must be clear to make the feature obvious. A small see-through feature would be ineffective. In addition, the design of this feature must take full advantage of the simultaneous front- and back-side offset printing capability in spot colors at the BEP so as to challenge the improving ink-jet printers' registration, color simulation, and resolution.

See-through registration is a visual feature in transmitted light, and although it could be read by a machine, its contribution is to create an overt visual feature primarily for the public and secondarily for the cash handler and teller.

A see-through registration feature can be found on many international currencies. Typically, the large numbers designating the denomination are split in half, with part of the number on one side and the remaining portion on the other. When held to the light, the entire number becomes visible. Because the public is very focused on the denomination of any particular note, there is a special focus on these numbers. Although see-through registration is not unique as a feature, as it is used in other countries, there is still room enough for innovative designs that it should remain effective against the primitive class of counterfeiter. The innovative designs could include more-complex patterns and the inclusion of subtractive color creations.

One possible registration feature would be to print a set of fine lines on both sides of the note that are precisely on top of each other. On a genuine note, the gaps between the lines would be clearly visible when the feature was held up to a light source. But on a counterfeit, in most instances, the variation in printer tolerances would result in the lines not being on top of each other, effectively closing the transmitted image gap, making the image look dark or solid.

Feature Motivation

With see-through registration offering an impressive combination of high saliency, easy implementability at the BEP, and the ability to deter the petty criminal, this feature is highly rated. In addition, this feature would have an immediate and positive impact on the public. Education of the public regarding this feature would be required, but a relatively simple explanation would suffice.

A carefully designed see-through registration feature could be an excellent and inexpensive way to deter the lower-level criminals from producing counterfeit notes. This feature can only be successful with the very precise alignment of the printing on the front and back sides of the FRN in spot colors. At present, there are no inexpensive two-sided printers available on the market, and there appears to be little motivation for creating this type of printer in the future beyond the environmental pressure to reduce paper usage by means of two-sided printing. The feature would have to be designed so that it was sophisticated enough to overcome the "good enough" registration possible on inexpensive ink-jet printers that counterfeiters have already used, according to data provided to the committee by the Secret Service.

The fine-line-feature approach might satisfy the basic requirements of being hard to replicate by the opportunist counterfeiter and easy to observe. However, of direct challenge to this feature class is the remarkably accurate print registration on the paper available with common, low-speed ink-jet printers.[10] Print registration is called for in a variety of special printing tasks that still rely on the generation of layers of artwork or graphics. Special printer drivers are being marketed for even dot-level control of the printing process with registration on paper of special material. Currently, the BEP has production processes in place that allow for the necessary degree of precise registration in spot colors, and therefore little to no cost would be incurred to implement this new feature.

Materials and Manufacturing Technology Options

The BEP already has the capability to control the precise location of the printed image on the substrate and, to the committee's knowledge, it appears that the BEP can obtain a level of precision that is necessary for creating the see-through registration feature.

[10]Some printers' photo ink-jet printer drivers add dithering algorithms to the printing on a page. It is not clear why this is done, but alternative drivers and work-arounds can be easily found.

There is a narrow design window for this feature type that will make it difficult to simulate using ink-jet printing. The design needs to take into account the pixelated and digitally addressable aspect of the ink-jet printer. The BEP equipment can avoid the image-processing methods of conventional printing and use or develop special security inks that are out of the normal color gamut to create a see-through feature that might fill this gap.

All ink-jet and commercial digital printing uses pixelated images with multiple dots partially overlapping to create the desired color. This process results in color lines that are wider than, or a subpixel offset compared with, a single ink-dot line. In addition, using special dyes in the offset ink can exacerbate this aspect of the simulated ink-jet-printed image. The unique security ink colors could also be used to BEP advantage in the same feature, to give a combined unique transmitted image color that is neither of the printed colors and will appear only with very high resolution single-pixilated straight-line printing. Again, the additive aspect of the spectrum will be washed out to some extent in wider, less-distinct-line ink-jet printing. An additional aspect is to combine see-through registration with a substrate feature such as the cream color of the FRN currency or a special dye added in the security thread.

Simulation Strategies

As most commercially available printers do not print two sides of a substrate simultaneously, precise registration requires some additional effort. Possibly only state-sponsored counterfeiters have access to the types of printers that could duplicate the registration precision required. Primitive and opportunist classes of counterfeiter would not be able to simulate this feature well enough to fool the general public. The petty criminal would have to obtain special inks and use specialized software drivers to ensure registration of the front and back sides of the note in separate printing. Also, using the multidimensional aspects such as incorporating a property of the substrate with this feature would force lamination or other measures to be used, similar to the simulation of other transmitted optical features by professional criminals.

Key Development Risks and Issues

Durability

With the degradation of the note over its lifetime, the transparency of the FRN may be obscured by the accumulation of dirt, which would limit the effectiveness of the see-through registration feature. However, it could be argued that the effectiveness of most features would be reduced by the accumulation of dirt.

Aesthetics

Using see-through registration as a feature would entail using a pattern large enough for observers to actually determine whether the registration was correct (authentic notes) or not aligned (counterfeit notes). Using a set of parallel lines or having only half of a number visible on one side (as is done in other international currencies) may appear too modern in design and potentially not appropriate for U.S. notes. However, there are many other ways to incorporate a registration feature by using other patterns that would be in keeping with the aesthetics of the note.

Social Acceptability

There are no issues of loss of privacy or environmental hazards related to the implementation of this feature.

Key Technical Challenges

The key technical challenge for the see-through registration feature would be the development of a design that took advantage of the nonpixelated, nondigitally addressable image-creation capability of the BEP using its offset spot-color printing, with the possible inclusion of a substrate color. The design would have to include precision in alignment or registration of the front and back sides of the note that is beyond the capability of desktop systems now and for the next 5 to 10 years. Although the machines at the BEP are capable of this level of precision, they are not used in this fashion at present. The stretching of the substrate in the presses could complicate registration. Also, the opacity of the paper may have to be adjusted to make this feature more visible in normal transmitted light.

Phase I Development Plan

Maturity of the Technology

Aspects of this feature are technically mature as they are in use in other currencies. However, to establish feature designs as described above would require standard manufacturing quality-control testing at the BEP. Some additional equipment would be needed for measuring the performance of the transmission aspect of this feature, and the BEP would have to test its machines for precise registration capability and determine the optimum level of paper opacity.

Current and Planned Related Developments

This feature design has some aspects in common with the high-complexity spatial patterns feature previously discussed. Development programs for this feature probably exist in other countries that use see-through registration. Because this feature is quite simple, the development of a similar program in the United States would require modest investment to implement.

Key Milestones

The milestones for implementing this feature are the following:

- Design a test see-through registration feature that is both pleasing to the eye and easy to authenticate and technologically challenging to the ink-jet printer, assuming continued improvement in resolution and printing registration.
- Test existing BEP printing equipment for precise registration and stability of colors.
- Modify the paper opacity specification as required.

Development Schedule

This feature is relatively simple. It is expected that it could be ready for production within 3 years.

Estimate of Production Cost

Compatibility with Current BEP Equipment and Processes

Adding the see-through registration feature would have minimal effect on manufacturing operations at the BEP. Only the level of precision at the BEP would need to be increased. It is possible that the scrap rate might increase.

Incremental Production Cost

The incremental production cost for this feature is estimated to be very low.

Required Capital Equipment

No major capital equipment would be necessary to produce this feature.

Further Reading

Alpern, M. 1978. The Eyes and Vision Handbook of Optics, W.G. Driscoll and W. Vaughan (eds.). NewYork: McGraw-Hill, Ch. 12.

Boucher, P.E. 1955. Fundamentals of Photography. New York: Van Nostrand.

Dowling, J., L. Stryer, and T. Wiesel. 1999. Vision: From photon to perception, pp. 1-3 in National Academy of Sciences Colloquium on Vision. Available at <http://www.nap.edu/catalog/9965.html>. Accessed February 2007.

Garcia, D.D. 2007. C What U C: A Visual Acuity Simulator. Available at <http://http.cs.berkeley.edu/optical/SPIE/SPIE98_CWhatUC.pdf>. Accessed February 2007.

Tassi, P., N. Pellerin, M. Moessinger, A. Hoeft, and A. Muzet. 2000. Visual resolution in humans fluctuates over the 24 hour period. Chronobiology International 17: 187-195.

Westheimer, G. 1981. Visual Hyperacuity: Progress in Sensory Physiology. Berlin: Springer-Verlag.

SUBWAVELENGTH OPTICAL DEVICES

Description

Conventional inks and printers offer powerful capabilities for accurately duplicating visible monochrome or color images, as evaluated by the unaided eye. Duplicating images that modulate the state of polarization or phase, or that have complex spectroscopic signatures, however, is much more difficult. Subwavelength optical elements (SOEs) provide visible images that achieve colors, polarization contrast, and/or other optical effects through the use of structures, such as relief features or narrow conducting wires that have dimensions substantially smaller than the wavelength of light. These optical characteristics, and especially those that provide polarization-dependent effects (for example, polarization-dependent transmission or reflection, or polarization transformation), offer a high level of control over the appearance of an image as viewed by the unaided eye under suitable lighting conditions. The subwavelength structures of SOEs can also provide forensic or machine-readable functionality.

The concept of an SOE feature is based on the wide range of optical characteristics that can be obtained from a single material by structuring it on length scales shorter than the wavelength of light—that is, substantially less than a few hundred nanometers. For example, parallel arrays of thick, subwavelength metal lines on a transparent substrate are transparent to light polarized along the lengths of the lines but do not transmit light with the orthogonal polarization. Similar structures in dielectric materials can rotate the polarization of the transmitted light, reflect light in narrow wavelength ranges, or serve as antireflection surfaces. Spatially variable structures of this type, used alone or in combination with conventional inks, can yield images with complex optical effects. For example, radially oriented subwavelength metal lines can produce polarizing elements for which the high-transmission direction has circular symmetry. This type of element would be difficult to construct using conventional polarizing optics. The chromatic effects, especially for the combined transmission, reflection and scattering effects, available in SOEs would be similarly difficult to simulate using conventional inks.

This feature is intended for use by the general public but is well suited also for machine readers. Of the various optical effects that are achievable with SOEs, the polarizing and polarization transforming operations appear to have particularly unique potential for the general public, provided that the public is educated on the conditions needed to view these effects—such as through polarizing sunglasses or with light from a backlit liquid-crystal display. The complex spectroscopic properties (for example, narrowband reflection) could also be useful in this sense, but in a manner that is most useful when combined with the assistance of another optical element or gadget—such as a narrowband filter—for viewing or in metameric patterns. These types of features, as well as the polarization effects, could easily be detected, at high speed, with a machine reader. The very small sizes of the structures in SOEs and the ability to incorporate nearly arbitrary geometries provide opportunities for functionality that could be useful for forensics.

Feature Motivation

The committee found the concept of the SOE feature to be useful because this feature has both overt functionality that can be used by the general public and by machine readers, and functionality for forensics. As a result, an SOE feature has potential value for protection against all classes of counterfeiter. Also, the manufacturing baseline and durability established by existing diffractive optically variable device (DOVD) currency features suggests technical feasibility for SOEs.

The main risk with this feature is that certain of its effects (for example, polarization response) are apparent only to an educated public. The manufacturing costs for fabrication at the required resolution must be established, although there is some promise for using scaled versions of the embossing techniques currently used for low-cost DOVDs such as holograms. The level of robustness of an SOE will be comparable to that of a DOVD.

SOEs are used in a variety of optical systems, and they are commercially available, currently for niche applications, from several vendors. The use of these items for applications in the visible or ultraviolet is described in the scientific and technical literature.

The committee is unaware of SOEs being used in currency or security applications. The structures in SOEs are similar to, in some cases, but smaller than those found in diffractive optical devices. The design of a particular type of SOE feature (that is, the colors, polarization behavior, and so on) would influence its uniqueness.

Materials and Manufacturing Technology Options

SOEs use established materials, structured into geometries that have subwavelength scales. Some of the manufacturing processes used to fabricate them rely on procedures borrowed from the microelectronics and display hologram industries. Newer techniques based on embossing procedures similar to those used in the production of blue semiconductor lasers for use in high-definition digital videodisc players are also suitable, as demonstrated in the recent scientific literature (Wang et al., 2006). The SOEs are similar in some ways to DOVDs, except that the sizes of the structures in SOEs are substantially smaller (2 to 10 times) than those in DOVDs. The SOEs can be integrated into currency in ways similar to those for DOVDs.

Simulation Strategies

Some of the color effects in SOEs could be simulated, for viewing with the unaided eye, using conventional inks. Simple polarization-dependent behaviors might be simulated by cutting and pasting conventional polarizers and waveplates onto a counterfeit banknote. These simulation methods are not, however, readily scalable for larger-scale counterfeit production. In addition, neither conventional inks nor cutting and pasting can reproduce many of the characteristics of SOEs—such as the circular polarizer configuration illustrated in Figure C-11. As a result, simulating SOEs with conventional inks would yield poor-quality simulations that could be identified by an educated public—that is, by a public that understands the viewing conditions needed to observe the features.

Strategies to duplicate SOEs that incorporate only surface relief structures would be similar to those used for surface-relief-type reflective or transmissive DOVDs. In this approach, the counterfeiter uses the optical element itself as an embossing tool to create duplicate relief structures with similar geometries, followed, in some cases, by coating with metal films by physical vapor deposition. The level of difficulty in using these methods with an SOE would be higher than the level of difficulty for a DOVD because the feature sizes are considerably smaller (for example, by 2 to 10 times). The effectiveness of the embossing approach could be reduced by including features in the SOE, such as patterns of isolated subwavelength metal lines, patterns of color, or other hybrid features (see the section above entitled "Hybrid Diffractive Optically Variable Devices") that could not be reproduced directly by embossing. The fabrication of such features requires sophisticated setups that are available only to professional criminal or state-sponsored counterfeiters.

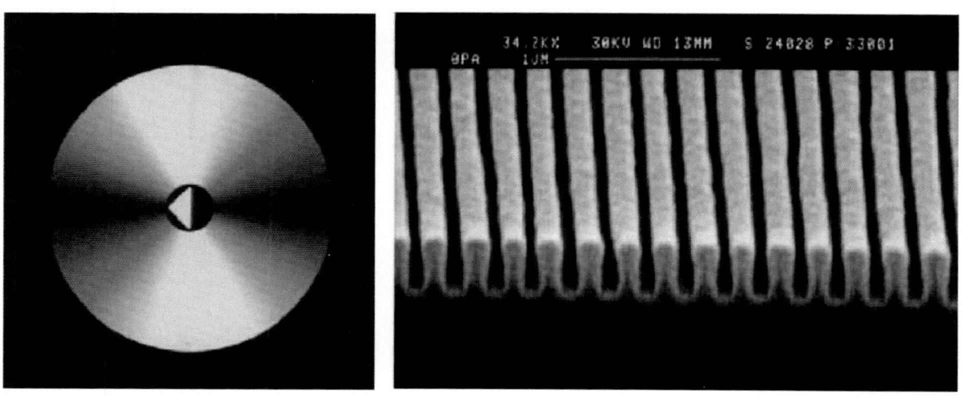

FIGURE C-11 Scanning electron micrograph (left) of a subwavelength optical device, which, in this case, consists of an array of metal lines. This structure forms a polarizer, whose implementation as a circular polarizing element is illustrated on the right. SOURCE: Schnabel et al. (1999).

Key Development Risks and Issues

Durability

The committee estimates that the durability of an SOE feature would be comparable with that of a diffractive optical device such as holograms or kinegrams. The degradation modes, for instance, would be similar.

Aesthetics

The specific design of an SOE feature determines its aesthetics. In some embodiments, an SOE feature would be aesthetically similar to a diffractive optical device. In others in which, for example, the SOE simply provides a polarization contrast, the SOE feature would alter only by a small amount the current look and feel of U.S. currency. In any case, the SOE would require integration with the note through a bonded plastic element similar to that used for a diffractive optical device.

Social Acceptability

The committee does not foresee any social acceptability issues with regard to this proposed feature.

Key Technical Challenges

The key technical question is whether the embossing-based fabrication approaches for SOEs, which are similar to those used for diffractive optical devices, but require higher resolution, can be scaled up for low-cost manufacturing. These methods are described in the scientific and technical literature, and they are in use by several companies for niche applications, primarily in the infrared or near-infrared.

Phase I Development Plan

Maturity of the Technology

SOEs have been described in multiple scientific and technical publications, and they form the basis of a set of niche commercial products, primarily for applications in the infrared and near-infrared. The basic materials and many aspects of the SOEs are similar to those of diffractive optical devices.

Current and Planned Related Developments

There are a number of research efforts at industrial and academic laboratories focused on the development of high-resolution embossing approaches for fabricating sub-100 nm structures for microelectronics and photonics. Many of these programs have been and are currently supported by agencies such as NSF and DARPA. For example, the NSF Center on Nanomanufacturing based at the University of California at Los Angeles and the University of California at Berkeley includes development efforts on related lithographic procedures based on embossing. These and similar methods represent the most promising path to the low-cost manufacture of SOEs.

Key Milestones

The key milestones for Phase I development are the following:

- Develop an approach to achieve high-volume production of SOEs for operation in the visible range.
- Develop suitable designs that balance aesthetics, security, and cost of manufacturing.
- Develop scanners and associated processing software to detect phase-encoded and other covert responses of SOEs.

Development Schedule

The committee estimates that the Phase I development for an SOE feature could be achieved with 2 to 4 years of development work. The key assumption is that the basic manufacturing approaches and materials used for SOEs that operate in the infrared and near-infrared will be suitable for SOEs in the visible.

Estimate of Production Cost

Compatibility with Current BEP Equipment and Processes

The SOEs could be integrated into a security strip or other component in a manner similar to that used for diffractive optical devices that are found in other currencies. The SOE would be integrated into the paper itself, thereby minimizing the effects on BEP manufacturing operations.

Incremental Production Cost

The cost to incorporate an SOE is estimated to be similar to that for a diffractive optical device.

Required Capital Equipment

New capital equipment would be required. It would be similar to that used for diffractive optical devices but with enhanced resolution capabilities.

References and Further Reading

Kikuta, H., H. Toyota, and W.J. Yu. 2003. Optical elements with subwavelength structured surfaces. Optical Review 10(2): 63-73.

Pelletier, V., K. Asakawa, M. Wu, D.H. Adamson, R.A. Register, and P.M. Chaikin. 2006. Aluminum nanowire polarizing grids: Fabrication and analysis. Applied Physics Letters 88: 211114.

Schnabel, B., E.-B. Kley, and F. Wyrowski. 1999. Study on polarizing visible light by subwavelength-period metal-stripe gratings. Optical Engineering 38(2): 220-226.

Wang, J.J., X. Deng, X. Liu, A. Nikolov, P. Sciortino, F. Liu, and L. Chen. 2006. Ultraviolet wave plates based on monolithic integration of two fully filled and planarized nanograting layers. Optics Letters 31(12): 1893-1895.

<http://www.nanoopto.com/>. Accessed March 2007.

<http://www.nanonex.com/>. Accessed March 2007.

<http://www.molecularimprints.com/>. Accessed March 2007.

TACTILE VARIANT SUBSTRATE

Description

The principal idea behind the proposed tactile variant feature on an FRN involves creating changes in the feel across the face of the bill that can be detected by the touch of a hand. The idea is that each denomination would have its own unique tactile signature.

To date, the only tactile feature on the FRN is the substrate itself. The special combination of cotton and linen combine to make a substrate that is not only unique, but very difficult to simulate. In general, any feature will be most salient if there is little "noise" in the "background." In this case, "noise" refers to other tactile features and "background" refers to the entire note. With the absence of other tactile features, the saliency of a new tactile variance feature would be even more pronounced. Introducing a tactile feature would involve roughening a specific area on the note in such a way that would make it "readable"—that is, easily detectable and discernible from the tactilely different areas on other denominations.

One example of a tactile variant feature could be vertical strips (in the same direction as the existing plastic ribbon) that are rougher than the base substrate. These strips could be of different widths, and they could vary in number, similar to a bar code (see Figure C-12). The challenge would be to choose a tactile variant pattern that would clearly indicate a certain denomination—for example, one thin strip for the $100 note, two thin strips for the $50, three thin strips for the $20, one medium strip for the $10, two medium strips for the $5, and one thick strip for the $1. The visually impaired would also be able to use the different denominations without confusion and with confidence. A teller could also use the feel of the currency to determine the authenticity of a note and potentially could use illumination from the side to inspect the tactile variant feature visually.

With the U.S. banknotes having a unique substrate of cotton and linen, creating small areas that are rougher to the touch would make the substrate even more unique and difficult to simulate. Also, with careful design of different tactile variant areas for each of the denominations, counterfeiters would find it difficult to reuse a bleached note of a smaller denomination to simulate a note of a larger denomination.

This feature is primarily a human-perceptible, not a machine-readable, feature. However, given the three-dimensional quality of a rougher substrate, it is conceivable that this feature would have visibility with illumination from the side of the note. A tactile variant note would also provide a unique opportunity for the blind and the visually impaired to recognize the denomination. A tactile variant feature would not require additional devices for its authentication or denomination; to the contrary, it would require a simple tactile test.

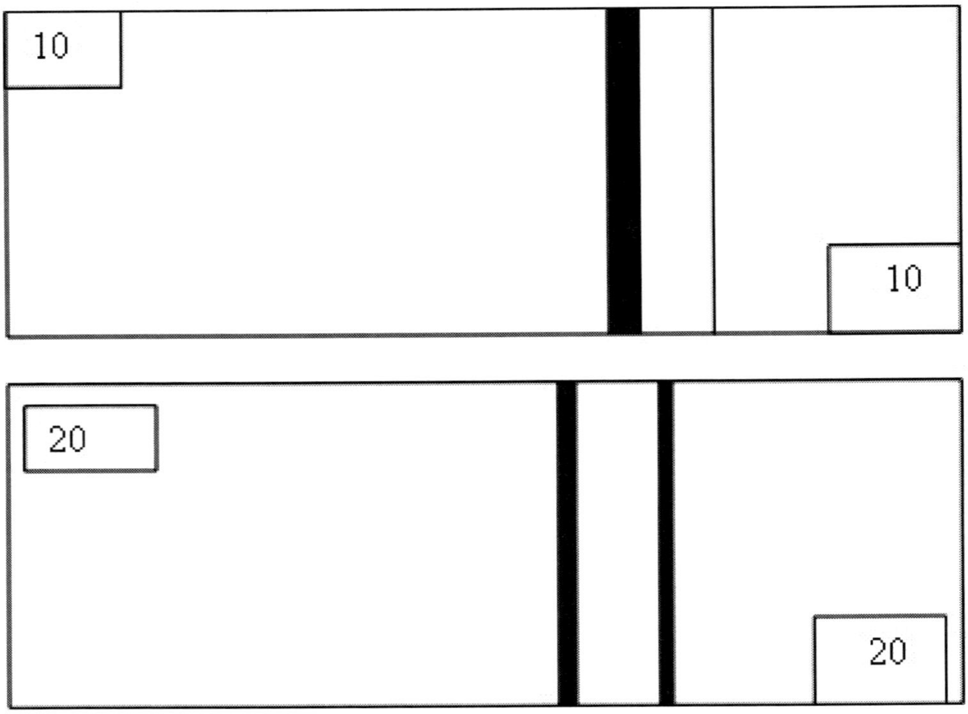

FIGURE C-12 Schematics of two denominations with different bar-code-like strips for the tactile identification of each.

Creating this feature would require the substrate manufacturer to add another step to the process. As paper-roughening machines already exist, it would appear that the cost to include this new feature would be minimal.

Feature Motivation

U.S. banknotes arguably have one of the most unique substrates of all the international currencies. Its feel is not only easily recognizable to the touch, but it is also extremely difficult to simulate. This special substrate has been and continues to be one of the most important features of the FRN. A tactile variant feature on the FRN is an excellent way to enhance the uniqueness of the cotton and linen substrate, while also creating a first-of-its-kind tactile feature. This feature would be particularly important because for the first time it would allow the visually impaired and the blind to differentiate between the various denominations.

A tactile variant feature allows the substrate of each denomination to be unique. As the quality of inexpensive printers continues to rise, the substrate is again becoming an important and critical feature. Not only would counterfeiters be less inclined to attempt to pass a fake note produced on poor-quality paper, but a tactile variant feature would make simulation even more difficult. Also, as indicates above, counterfeiters have been known to bleach FRNs in order to use the substrate to make counterfeits of higher denominations, and this new feature would reduce such counterfeiting considerably. This feature would probably be well received by the public, which is already very aware of the feel of the note.

Materials and Manufacturing Technology Options

To create a tactile variant feature, a machine process would have to be created at the manufacturing facility of the substrate supplier. The machine would have an abrasive quality that would precisely roughen up the surface in specific areas. The BEP would receive the substrate with this feature already added. The feature could, for instance, be added to the side of the portrait, in an area where other features would not be adversely affected.

Simulation Strategies

Simulation of this feature is possible in a crude fashion—that is, using a substrate-roughening machine. However, because the careful design of a tactile feature would not allow a bleached lower denomination to be used to simulate a high-denomination note, simulation would become very difficult. Thus, this feature would be effective in deterring the primitive, opportunist, and petty criminal, and to some extent the professional criminal.

Key Development Risks and Issues

Durability

The tactile qualities of the substrate could potentially change over the course of the note's lifetime, with the rough areas (feature) becoming softer over time. However, the main characteristic of this feature is that the rougher areas remain distinct and different from the normal tactile qualities of the substrate. It is the committee's determination that this difference in roughness would remain over the note's lifetime, although durability experiments would be needed to confirm this.

Aesthetics

Using a tactile variant feature would not degrade the feel of the substrate; in fact, it might even bring more attention to the tactile quality of the FRN. Placement of the feature could potentially degrade or distort the look of the portrait; thus, it perhaps would be prudent to place this feature to the right of the portrait.

Social Acceptability

There are no obvious hazards or public concerns created by introducing this feature. Although a simple bar-code design is suggested, this design would be very rudimentary (only a few strips) and therefore could not contain the amount of information that a normal bar code could. Thus, there would not be any loss of privacy. There are no environmental hazards. The inclusion of a feature that the visually impaired and blind population could use to denominate and authenticate FRNs would be of great social benefit.

Key Technical Challenges

The technical challenges for the tactile variant feature involve establishing the process by which the substrate becomes rough and understanding the manner in which the ink will respond to a substrate with varying levels of roughness. Adding roughened areas to FRNs needs to be done in a such a way that whole sheets of bills can still be printed. The use of a bar-code-like design has its benefits, as one could apply roughened strips straight across the entire sheet. This could be the simplest format, thereby using a roughening machine directly on the roll of paper as it exits the papermaking machine—that is, before it is cut into sheets. Using roughened shapes that repeat themselves along an imaginary vertical line is another approach to implementing this feature. High-pressure printing on a substrate with varying roughness will probably not cause any adverse effects to the precision and clarity of the note. Experiments will be needed to confirm this.

Phase I Development Plan

Maturity of the Technology

Experiments for durability and ink clarity need to be designed and performed, a design for this feature needs to be carefully chosen, and a machine to create the unique rough areas on the FRN needs to be created or modified from an existing one.

Current and Planned Related Developments

There are no known development programs related to this feature, but the feature is quite simple, and it is expected that relatively little investment would be needed to implement it.

Key Milestones

The milestones for the Phase I development of this feature are the following:

- Conduct a laboratory demonstration of a substrate-roughening method that can produce roughened areas of different shapes.
- Develop test features incorporating tactile variance.
- Carry out experiments to estimate the durability of a roughened substrate.
- Conduct initial experiments to determine how intaglio images would be affected by a substrate with different degrees of roughness.

Development Schedule and Cost Estimate

It is expected that the Phase I, and indeed perhaps the total development, of this feature could be completed within 2 to 3 years.

Estimate of Production Cost

Compatibility with Current BEP Equipment and Processes

The roughening machine to create this feature would be used by and located at the supplier of the substrate. Thus, the effect on BEP operations would be minimal.

Incremental Production Cost

The cost for the roughening machine should be very modest, and it is expected there would be a low incremental cost.

Required Capital Equipment

The technology to create a roughening machine already exists. In fact, these machines themselves already exist. They would, however, need to be modified so that they are capable of creating the unique feature shapes on the FRN.

Further Reading

No additional reading is suggested.

THERMORESPONSIVE OPTICALLY VARIABLE DEVICES

Description

A thermoresponsive optically variable device (TOVD) provides an appearance that changes reversibly with temperature—for example, blue when cool and red when hot—over a temperature range near room temperature. Heating this type of responsive device (or cooling it, depending on its temperature) by touching it with a finger, for example, changes its appearance. In certain embodiments, such as those that involve liquid crystals, the colors can also change with viewing angle, and the properties of the reflected or transmitted light can depend on polarization. Such devices would be difficult or impossible to reproduce with conventional scanner or printer technologies. They provide additional functionality—that is, responsiveness—and are challenging to simulate or duplicate, compared with conventional OVI features or DOVDs. As such, they provide enhanced security benefits. Their highly visible, reversible, and optically variable appearance and the polarized nature of the reflected or transmitted light provide a level of banknote feature functionality that is valuable for the general public.

One type of TOVD can be formed with thermotropic chiral liquid-crystal (LC) materials. The nematic phase of a chiral LC is known as the cholesteric phase, and is observed in LCs with chiral nature or in achiral LCs that have chiral additives. This phase consists of a helical arrangement of LC molecules, with a well-defined pitch. Circularly polarized light that strikes a layer of cholesteric LC is reflected when its wavelength is comparable to the distance associated with a full turn of the helix (that is, the pitch of the cholesteric phase). This effect causes the LC to appear brightly colored. The color depends on the viewing angle owing to the geometry of the Bragg effects that generate the reflected light. The reflected light also can be circularly polarized, providing additional benefits for security. The color varies with temperature owing to the temperature dependence of the helical pitch of LC molecules in the cholesteric phase. The chemical structure of the LC molecules determines the pitch and its dependence on temperature.

Figure C-13 illustrates the thermal effects in a typical system. A TOVD currency feature could consist of a uniform patch of this type of material, a printed image formed on top of such a patch, or an image formed directly with the thermoresponsive material. Separate fabrication of the feature followed by integration with the paper represents a path to insertion into currency. TOVDs based on liquid crystals are used in thermometers, mood rings, car paints, and battery testers. They are also being explored for use in low-cost systems, such as "re-printable" paper. These devices can be used in security applications, but the committee is unaware of any widespread use for such purposes. The main challenges for currency applications are durability and cost.

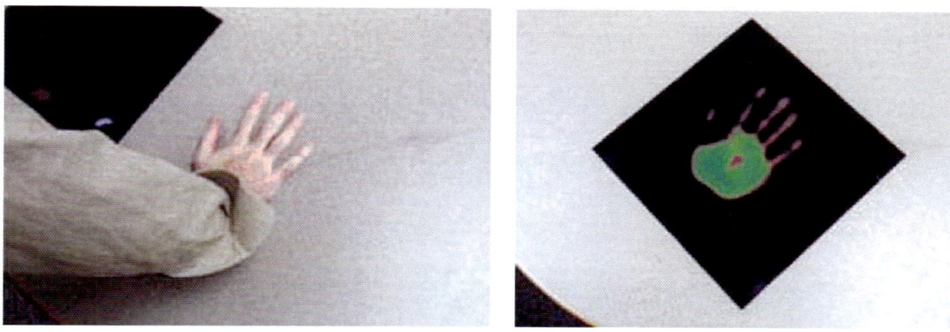

FIGURE C-13 Images showing the thermal response of a cholesteric liquid crystal. At room temperature, the device is black (upper left in the left frame). Heating by a hand causes changes in the observed color (right frame). SOURCE: Adapted from the University of Wisconsin-Madison IPSE Liquid Crystal Activity Guide <http://mrsec.wisc.edu/Edetc/supplies/ActivityGuides/LC_Activity_Guide_Expo.pdf>, Materials Research Science and Engineering Center on Nanostructured Interfaces, University of Wisconsin-Madison.

Another route to TOVDs uses thermoresponsive inks known as leucodyes. Although these materials are much better explored for security applications than are liquid-crystal-based devices, they have the disadvantage that they do not provide unique viewing-angle and polarization-dependent properties. They are, however, more fully developed for low-cost implementations and are used widely in packaging applications. Durability represents the main challenge for currency applications.

Feature Motivation

The committee considered the concept of a TOVD feature to be valuable owing to its easily identifiable, unique, and highly visible responsive functionality, suitable for use by the general public even with little education provided on the nature of the feature. It is also usable by high-speed machine readers. A TOVD feature would provide, however, limited forensic functionality.

As described above, heating or cooling a TOVD feature by touching it with a finger, for example, changes its appearance. Such a characteristic would be difficult to reproduce with conventional scanner or printer technologies. Implementations with cholesteric liquid crystals offer the widest diversity of visual indicators, including viewing-angle-dependent appearance and polarization-selective operation. Leucodyes, however, are more fully developed for low-cost applications. In both cases, robustness for currency applications must be demonstrated.

Materials and Manufacturing Technology Options

A TOVD requires a thermoresponsive material (for example, thermotropic chiral LCs or leucodyes), a suitable substrate, and a top layer to seal the system. The manufacturing processes for cholesteric and leucodye devices are well established. Specially designed materials, especially in the case of the LCs, could provide unique colors and temperature responses. Such a device could be formed into an element, such as a security strip, and then integrated with the paper prior to printing. The ability of such a structure to withstand the direct pressure of the intaglio process is unknown. It might be possible, however, to adjust the position to avoid direct contact during printing or to integrate the TOVD after printing.

The designs would need to take into account a wide range of operating temperatures—for use, for example, in regions from Alaska to Arizona. However, the functioning of the devices—that is, their responsiveness—only requires changes in temperature due to contact with a finger or other device. As a result, the ambient temperature does not limit the device's operation except in the special case that the temperature of the note is the same as body temperature, as in the case of inspection of the feature with a finger. To expand the complexity of the functioning of a TOVD and to address the variable ambient temperatures, it would be possible to construct a feature that consisted of a composite array of different features, each of which responds in different temperature ranges.

Simulation Strategies

For liquid-crystal-based devices, simulation might be possible at a crude level, by cutting and pasting off-the-shelf devices obtained from battery testers, decorative items, thermometers, clothing, and so on. TOVD features that involve printed patterns, or their integration with other printed images, would make this sort of simulation strategy difficult. Also, the colors and temperature responses of TOVDs used in currency could be custom-designed to differ in identifiable ways from commercially available devices. Nevertheless, it is reasonable to expect that crude simulations could be generated by petty criminals and that acceptable simulations could be produced by professional criminals. Duplication of TOVDs, while more difficult than simulation, is within the range of capabilities of a state-sponsored organization. Devices based on leucodyes might be easier to simulate, owing to the wider availability of the inks as well as the feasibility of directly printing patterns of leucodyes without concern for careful control of thickness or molecular alignment, which are necessary in the case of liquid crystals. A path to simulation of a liquid-crystal device could use, in fact, combinations of leucodyes and conventional pigments to simulate the color changes. The viewing angle and polarization-dependent behavior would, however, be difficult to capture using this approach.

Key Development Risks and Issues

Durability

The durability of TOVDs is sufficient for use in a range of existing applications, as noted above, but their durability, at the level required for currency, is unproven. The degradation modes would be a function mainly of the device packaging, particularly in the case of the liquid-crystal systems, since the materials themselves are known to be very stable—liquid-crystal display applications provide an example. The leucodyes can be degraded by prolonged exposure to ultraviolet light. Suitable packages (for example, ultraviolet-absorbing encapsulation layers) would need to be developed to avoid these sorts of limitations.

Aesthetics

A responsive feature, such as a TOVD, could, if properly incorporated, enhance the aesthetics and appeal of U.S. currency.

Social Acceptability

There are no known environmental hazards or public concerns with respect to TOVDs.

Key Technical Challenges

A key challenge would be the development of low-cost, high-volume manufacturing approaches for TOVD production. The extremely high levels of reliability and durability demanded by currency applications—including in this instance the range of ambient temperatures in which a thermal device would have to operate effectively—represent the main difficulty. Tests must be performed to assess the durability of existing devices for use in currency. The outcome of such tests can provide guidance on the development of suitable packaging systems. Methods to reduce the cost of the liquid-crystal-based devices, in particular, appear necessary.

Phase I Development Plan

Maturity of the Technology

The existing TOVD devices in the applications noted above suggest that the technology is mature for applications similar to but with less stringent operational demands than those in currency. These devices have not been demonstrated to achieve the necessary durability and reliability for currency applications, to the committee's knowledge.

Current and Planned Related Developments

There are many development efforts in liquid crystals generally, and in cholesteric liquid crystals in particular, to support existing product applications and to develop new systems, such as bistable reflective displays and "re-printable" paper, that use these materials. Additional efforts are required, however, to address the technical needs of the currency application, and in particular the cost.

Key Milestones

The key milestones to the development of a TOVD feature are the following:

- Conduct initial durability tests of several different TOVD features.
- Develop a reasonable approach to achieving cost-effective integration of TOVD features in the currency substrate.

Development Schedule

The committee estimates the time for completion of Phase I of development for a TOVD feature to be within 2 to 3 years. Successful systems-level tests would require, primarily, adequate packaging of the feature and cost-effective manufacturing approaches. The key assumption is that durability of a TOVD can be achieved by suitable packaging and integration approaches.

Estimate of Production Cost

Compatibility with Current BEP Equipment and Processes

The cost of a liquid-crystal-based TOVD is expected to be relatively high compared with that of other nonresponsive complex features such as diffractive optical devices. Integration of a liquid-crystal TOVD would occur through a strip bonded to or woven into the paper. TOVDs that use microencapsulated leucodyes might be printed directly. The layout of the printed parts of the currency might need to be designed to avoid high-pressure contact with the TOVD associated with the printing. Alternatively, development efforts could be directed to yield TOVDs that are compatible with these pressures.

Incremental Production Cost

The cost of a liquid-crystal TOVD would likely be in the medium to high range. Other approaches, such as those based on leucodyes, would be lower in cost.

Required Capital Equipment

Capital equipment used for existing TOVDs could be implemented directly, or in some variants, for currency applications. Liquid-crystal TOVDs would most easily be integrated through a security strip or on a plastic substrate that is integrated with the paper note, similar to a diffractive optical device. Leucodyes could be printed directly, although studies would be needed to determine whether existing BEP printers could be used for this purpose.

Further Reading

For more information on color-changing inks, see <http://www.screenweb.com/inks/cont/brighten981119.html>. Accessed February 2007.

Bahadur, B. 1998. Liquid Crystal Applications and Uses, Vols. 1-3. Singapore: World Scientific.

Broan, L., and C.L. Saluja. 1978. The use of cholesteric liquid crystals for surface temperature visualization of film cooling processes. Journal of Physics E: Scientific Instrumentation 11: 1068-1072.

Ireland, P.T., and T.V. Jones. 2000. Liquid crystal measurements of heat transfer and surface shear stress. Measurement Science and Technology 11: 969-986.

Parker, R. 1988. Flexible Resistive Heat Battery Tester and Holder. U.S. Patent 4726661.

WINDOW

Description

The principle of the window feature is the inclusion of a denomination-specific window, possibly integrated into the substrate of U.S. banknotes. The window could be shaped differently on different notes, or different notes could simply have windows of different sizes—in particular it is suggested that the lower-denomination notes have larger windows, thereby deterring their use for counterfeiting larger-denomination notes that would have smaller windows. Alternatively, higher-denomination notes could have no window.

Feature Motivation

The motivation for the window feature is its effectiveness in deterring the use of lower-denomination notes as sources of currency substrate for the counterfeiting of larger-denomination notes by the inclusion of a denomination-specific window or hole in the note. This feature is intended for unassisted use by the general public—that is, a large-denomination note would have a window, or the wrong window, if it was a counterfeit manufactured by the "washing" of a lower-denomination note.

In particular, therefore, the feature is proposed as a specific deterrent for the opportunist and petty criminal classes of counterfeiter. Also, the committee believes that the careful design of the window could add to this feature's value as a denominating aid for the blind.

The major risk with this feature is that robust embedding of a plastic window into the paper substrate might be difficult. This feature is expected to be cost-effective once the manufacturing challenge is solved, since the cost of the mass-produced plastic windows is expected to be less than that of the security thread. While clear plastic windows are in use in banknotes with plastic substrates—such as those in Australia and Mexico—the committee is not aware of integrated plastic windows in paper notes.

Materials and Manufacturing Technology Options

Substrate-integration technology similar to that needed to implement windows is also needed for the Fresnel lens, hybrid diffractive optically variable device, microoptic array, and subwavelength optical element features discussed earlier in this appendix. In order to produce this feature so that it is challenging to the counterfeiter, the window would have to be integrated into the substrate. This would require a change in the substrate manufacturing process, perhaps as a variation of the process used to integrate the current security strip.

Simulation Strategies

The windows could be simulated and duplicated by professional criminal and state-sponsored counterfeiters in counterfeiter-produced substrates. However, simulation by a petty or opportunist counterfeiter would be challenging, as it would require the holes in the lower-denomination notes to be "filled in" in order to allow for the use of those notes by bleaching for instance, as a substrate for higher-denomination notes.

Key Development Risks and Issues

Durability

The positioning, size, and shape of the window would have to be investigated to minimize durability issues. The committee believes that given the remarkable strength of U.S. currency paper, the durability of the feature and the note has a high probability of being satisfactory.

Aesthetics

The feature will change the aesthetics of the U.S. FRN, but properly designed, the window should not affect the overall look and feel of the notes.

Social Acceptability

There should be no social acceptability issues involved with the implementation of the window feature.

Key Technical Challenges

The key challenge is the development of the window design and production process that are respectively durable and inexpensive.

Phase I Development Plan

Maturity of the Technology

The window feature is a low-technology feature that could be implemented in the short term once the durability and production cost issues were resolved.

Current and Planned Related Developments

The committee is unaware of any research specifically targeting this type of feature. The committee knows of no plastic films that have yet been integrated into a paper substrate, although there does not appear any compelling reason to think that this cannot be done.

Key Milestones

The key milestones for the Phase I development of this feature are the following:

- Demonstrate that a plastic window can be effectively integrated into the substrate.
- Resolve durability issues.

Development Schedule

The committee believes that this feature could be fully developed within 2 years.

Estimate of Production Costs

Compatibility with Current BEP Equipment and Processes

The window feature might require an additional processing step in the production at the BEP, depending on whether the holes are produced prior to printing (by the substrate manufacturer) or after printing (at the BEP).

Incremental Production Cost

The committee believes that this feature would not add significantly to the cost of FRN production.

Required Capital Equipment

The need for extra equipment at the BEP would depend on whether the holes were produced prior to printing (by the substrate manufacturer) or after printing (at the BEP).

Further Reading

No additional reading is suggested.

D

Long-Term Feature and Feature Platform Descriptions

This appendix has in-depth feature descriptions of the longer-term features and feature platforms—that is, features that can be implemented in a time frame of more than 7 years. These features are also discussed in Chapter 5. Each feature description includes subheadings dealing with various aspects of the feature:

- *Description*—An explanation of the physical principle(s) on which the feature is based. Also, the feature application as visible, machine-readable, applicable to the visually impaired, forensic applicability, and so on is described. Furthermore, the benefits and limitations of the feature are presented; graphics may be included to depict the feature and its operation.
- *Feature Motivation*—A summary of the reasons why the feature is highly rated by the committee and reference to its uniqueness.
- *Potential Implementations*—A description of scenarios that provide examples of how the feature could be employed to deter counterfeiting.
- *Materials and Manufacturing Technology Options*—A summary of the materials and manufacturing process that could be used to produce the feature as well as initial thoughts on how the feature could be integrated into a Federal Reserve note.
- *Simulation Strategies*—A discussion of potential ways in which a counterfeiter could simulate or duplicate the feature, and the expected degree of difficulty in attempting to do so.
- *Key Development Risks and Issues: Phase I*—A discussion of the durability challenges, feature aesthetics, anticipated social acceptability, and descrip-

tion of the key technical challenges that must be addressed during the first phase of the development process to demonstrate the feasibility of the feature idea: that is, demonstrate feature capabilities and determine the usefulness in counterfeit deterrence. (The development phases are defined in Chapter 6.)

- *Development Plan: Phase I*—A characterization of the current maturation level of the feature technology, key milestones to be achieved during Phase I, known current and planned related developments external to the Bureau of Engraving and Printing (BEP), and a high-level schedule for Phase I.
- *Estimate of Implemented Production Costs*—An initial assessment of additional BEP operational steps that would be required at the BEP to produce a banknote with the feature, incremental cost (higher, lower, the same) relative to the cost of the current security thread, and an indication of whether additional BEP capital equipment would be required for production.
- *References and Further Reading*—Selected references relating to the feature and its associated components. Such references could include, for example, papers and conference proceedings for background on any work done relating to this feature. These lists are not exhaustive but are intended to provide a snapshot of current work related to the feature concept.

The features described in this appendix are as follows:

- Anomalous Currency Space
- Chemical Sensors
- Digitally Encrypted Substrate
- Engineered Cotton Fibers
- e-Substrate
- NiTi Shape Memory and Superelastic Responsive Materials
- Smart Nanomaterials
- Tactilely Active Electronic Features

ANOMALOUS CURRENCY SPACE

Description

The "anomalous currency space," or ACS (pronounced "ace"), is a materials-based approach that can serve as a platform for a wide range of anticounterfeiting strategies by providing a region or regions that differ entirely in materials composition from the banknote substrate. The primary objective is to provide an eye-catching visual feature that would also possess tactile properties. It is expected that the unique structure and composition of advanced materials incorporated into an ACS would assist forensic investigation as well.

This empty space can notionally be thought of as a clear plastic window. However, it does not have to be shaped like a typical window. For example, the shape could be a strip that runs the full length or width of the banknote, or a strip that runs along any or all of the edges of the note, or a series of regions dispersed throughout the currency note. Also, the materials composition of this region does not necessarily need to be a clear plastic or other polymer.

In this context, "anomalous" is used to emphasize that there is a physical space within the banknote that differs dramatically in terms of materials composition and behavior from the rest of the bill. Because Federal Reserve notes (FRNs) and their analogues already use a multiplicity of features that differ dramatically in materials structure and properties, the term "anomalous" refers to a macroscopic region of radical discontinuity relative to the bulk composition of the bill or note.

Feature Motivation

This feature platform offers numerous ways to create an eye-catching visual feature that could also possess other properties, such as a distinctive feel. The ACS provides for the incorporation of heterogeneous materials into the FRN in a manner that would not allow the ACS region to be inconspicuously removed or tampered with; also, these materials would have durability for the lifetime of the banknote.

Potential Implementations

The clear plastic window or variations on this theme are already in use in some foreign currencies, so it appears that this feature platform concept has been successfully introduced. The concept here is to extend the scope of this feature significantly. The polymeric material itself may be further modified in any number of ways, including the following:

- Direct integration of electro-optical or other types of materials within the polymer. The material could be anything up to and including complete integrated circuitry.
- Surface etching or other physicochemical modifications to one or both surfaces (back and front) of the ACS feature to create novel electro-optical or other effects.
- Complete perforation of the polymer to enable a number of effects, from patterned microholes for physical identification to a type of diffraction grating.
- Incorporation of "smart" materials (including nanomaterials) whose unusual properties are based on new compositions and/or structures and that are capable of dynamic interaction with the environment: for example, memory polymers that shape-shift on the basis of a change in a physicochemical parameter (for example, temperature).
- Use of dumb materials (including nanomaterials) whose unusual properties are based on new types of composition and/or structure but do not respond to environmental stimuli: for example, an ultratough polymeric strip that traverses the entire border of the FRN and is impossible to tear.
- Inclusion of other materials in or on the polymer to create composites with various properties (for example, holographic metal strips).
- Employment of other materials with unique active and/or passive properties, structures, or behaviors.

The window could be composed of two (or more) layers. Polymeric layers, for example, could have any or all of the properties described above. Further, the space(s) between the polymeric layers could contain additional materials. In the simplest case, two polymeric layers would be embedded flush with the two surfaces of the FRN, with each layer being less than 50 percent of the total thickness of the note itself. For example, assume a thickness of ~100 micrometers for U.S. currency and a thickness of 40 micrometers for a single polymeric layer. Assuming the polymer layers do not collapse and adhere to each other, the linear space between them is 20 micrometers in the center region (the ends would be embedded into the substrate). If the "window" is 1 centimeter square, the three-dimensional space between the two polymeric layers creates a volume of 2 microliters. This volume could be (fully or partially) filled with a novel material or composite, including current microscale and nanoscale materials and those in development as part of the National Nanotechnology Initiative (NNI). These materials and composites could be smart or dumb, active or passive, and so on. A larger window or thinner polymer layer would increase the available interlayer volume.

Given that U.S. currency exhibits desirable characteristics of materials strength, toughness, and so on with a thickness of ~100 micrometers, it is not unreasonable to assume the existence of materials that provide the same level of performance at less than half that thickness which, in turn, creates the interlayer space discussed about. Since future research will be conducted on supertough materials, it is reasonable to assume that layer thicknesses down to 10 micrometers or even thinner might be achievable. Such thinness would create the opportunity for larger interlayer volumes and/or multiple layers within the window. In the latter case, holographic-like effects and/or color-changing effects should be possible.

There are many ways in which a banknote designer could apply the ACS feature platform concept. A few ideas include the following:

- *Strip along outer edge.* A strip of ultratough materials around the outer edge of the bill could be easily detected by its distinct look and feel—this region could not be torn or perforated.
- *Distribution of small ACS features throughout the banknote.* Distribution of smaller ACSs throughout the bill could produce easily recognizable patterns that could be used for visible, tactile, and possibly instrument-based detection. Such patterns could exhibit dynamic as well as passive behavior if, for example, memory metals or polymers were used.
- *Memory polymers.* Memory materials fall into a larger category of smart materials that exhibit unique behavior. Memory polymers (often constructed of bulk copolymers) are capable of dynamic movement and associated shape-shifting when the variable of state is applied (often a change in temperature). For example, spatial distribution of a memory polymer within a certain space could allow the surface to change when it fell above or below body temperature. A simple case would involve reversible surface stippling that would manifest as a change in roughness, which could be detected qualitatively by touch and quantified by instrumentation. Embedded circuitry is specifically excluded from consideration here, so the behavior of these types of smart materials would be strictly dependent on the composition of the material itself and would require no dedicated power source.

As a further example, memory polymers have been produced for several biomedical applications, including self-tying surgical sutures. A memory polymer can exist in either of two states: elongated (two-dimensional fiber) or contracted (three-dimensional cylindrical). This material, produced by block copolymerization, could be incorporated into a clear plastic window in such a manner that the cylindrical watermark-like image could be created by the variation in light transmission produced by the three-dimensional state. Suppose the transition temperature was adjusted to ~35°C. The simulated watermark would appear in the window at temperatures below 35°C and disappear above 35°C. It is reasonable to suppose that U.S. currency spends more than 90 percent of its time at temperatures below 35°C. Therefore, FRN verification could occur by vigorously rubbing the window between the thumb and forefinger. Because the window would be ≥100 micrometers thick, frictional heating would rapidly raise the temperature, causing the image to disappear. As soon as the window was allowed to cool, the image would reappear.

A simpler alternative might be to construct the window so that the image disappeared when the window was physically stretched in one (or any) direction relative to the xyz coordinates of the FRN itself. This could be accomplished by controlling the orientation of the coiled memory polymers during fabrication so that the image was formed much like crosshatched pen and ink work or the engraving process itself. Since the image would disappear rather than becoming distorted, this trait could not be simulated by simply using a window made of an elastic material.

Materials and Manufacturing Technology Options

The ACS, by definition, has no specific material requirement other than being a completely different material from the bulk of the FRN. The requisite manufacturing technologies would depend entirely on the feature design and the materials selected for the feature application. The ACS provides a flexible mechanism whereby these advanced materials can be incorporated into currency with minimum disruption to the production process. Like the silicon wafer facilities required for integrated-circuit fabrication, extremely-high-technology equipment would be required for the production of some of these new materials, but once the manufacturing process was online the cost per unit would drop to the level of a bulk commodity. In other cases, the properties will depend on exact nanofabrication that will not be possible to counterfeit or simulate without complete knowledge of the molecular structures of the components, once again putting counterfeiting out of the reach of all but the most sophisticated criminals. The spectrum of physicochemical properties that could be incorporated into the ACS is as large as the spectrum of 21st-century materials, which means that the unique property could be physical, chemical, optical, electromagnetic, and so on.

Simulation Strategies

The ability to simulate an ACS feature will depend on the behavior of the material(s) selected. It is assumed that the wide range of materials that are under development will make it possible to select material performance traits that will make simulation extremely difficult, requiring resources above the level of the opportunist counterfeiter.

Using the example of the shape-shifting memory polymer: temperature-sensitive, reversible changes in surface roughness would be extremely difficult to simulate. The reason is that the properties of the feature are based on both the unique composition of the material and the method in which it is integrated into the ACS. Likewise, the ultratough, flexible, lightweight materials currently under development in places such as the Massachusetts Institute of Technology's Institute for Soldier Nanotechnologies cannot be simulated, since they are new materials with novel properties that are dependent on yet-to-be-created processing methods. It is highly probable that many of these methods will require a very high initial investment in sophisticated instrumentation. Therefore, initially there will be no analogues available for simulation and (for properties such as ultrahigh tensile strength) no way to simulate them.

Key Development Risks and Issues: Phase I

Since a wide range of highly durable materials will be available for the ACS, as well as rigorous testing methods by which this durability may be characterized, durability will need to be evaluated but probably will not be a major constraint. The ability to integrate a window or other ACS into the FRN is really a question of whether sufficiently strong bonding can be formed between the cotton-linen paper and the material(s) used to create the ACS. Given current and future methods for creating materials composites, it is highly probable that the ACS can be integrated into the FRN in a secure and durable manner, but the details would be part of the development program.

In terms of aesthetics, many advanced materials could provide a high-technology gloss to the FRN that would make it appealing to many users. Paper currency that possesses a region that changes shape or contains a perimeter that cannot be ripped, cut with a knife, or even perforated with a bullet would probably be received favorably by most of the public. By specifically limiting any new property to an ACS, the Bureau of Engraving and Printing (BEP) can retain most of the traditional look of the "American greenback." Retention of the traditional craft involved in creating U.S. currency and the reliability that has accrued to the "brand identification" are, in and of themselves, highly desirable aesthetically.

Another key implementation consideration involves the selection of the materials to ensure that they would not pose an environmental or health hazard. Great care will need to be exercised in the selection of any advanced material for the ACS. One simple yet profound example is an ACS capable of degrading into its component nanoparticles. Little is known about the toxicology of nanoparticles. Further, any specific nanofabricated material will have its own set of physicochemical properties. An ACS might meet with all reasonable durability standards for use yet, when burned, release nanoparticulates that could be inhaled (say by a child). Until there is a great deal more experience with these types of materials, the potential to set off an irrational panic response based on the incorporation of any new, advanced material (nanotechnology-based or otherwise) must be considered.

The specific key technical challenges with respect to the ACS will depend entirely on the materials selected for it. However, one of the advantages of choosing an ACS component for currency is that many key technical challenges for the materials themselves will be driven by other extremely-high-value applications such as medical implants or biodefense. As a result, most of the research and development (R&D) on the materials themselves can be leveraged by the BEP. Certain technical challenges will be intrinsic to the specific use of these advanced materials in FRNs. The most obvious is the physical incorporation of the ACS into the FRN.

Can *any* ACS feature be physically incorporated into the FRN? Using the clear plastic window as a simple example, it is reasonable to assume that if the material used in the ACS can be fabricated to conform with the physical dimensions of the FRN, then it may be incorporated into the FRN through some form of compositing. It is more likely that the key decision will be the cost of integrating any such compositing step into the current FRN production process. The BEP already has significant experience in incorporating anomalous materials into specific regions of the FRN (for example, specific regions of color-shifting ink and the security thread). In many ways, the ACS may be viewed as a logical extension of such features.

Development Plan: Phase I

Activity during Phase I must address the key technical challenges with respect to the use of the ACS in U.S. banknotes. There are many opportunities for features within this technology area, so Phase I activities can determine which potential features are of greatest interest for counterfeit deterrence. Feasibility experiments can be conducted, through a combination of laboratory-based work and modeling and simulation analysis, to select the most promising future directions for currency applications.

A thorough review of current related work would be the first priority. Several important development programs can be leveraged to create ACS-based features rapidly. This materials revolution is expected to produce new materials with properties that could be of tremendous usefulness in anticounterfeiting efforts. The NNI has already been mentioned. Specifically, the Department of the Treasury is already a member of the Nano-scale Science, Engineering and Technology (NSET) Subcommittee of the National Science and Technology Council that provides a mechanism by which the BEP can obtain information about candidate materials for an ACS. In addition to or in coordination with the NNI, advanced materials are under development at all major government agencies funded by an equally wide range of missions. Obvious candidate agencies would be the Department of

Defense (for example, the Institute for Soldier Nanotechnologies), NASA, and the basic sciences division of the Department of Energy.

Since there are so many possibilities of new features, multiple feature concepts can be pursued. Therefore, there could be a base level of activity that is continually at work in Phase I to further develop new ACS ideas. Then, specific ideas that have been determined to have attractive counterfeit-deterrence benefits could be spun out of the base program into distinct, defined projects that would proceed through all the requisite development steps.

Estimate of Implemented Production Cost

Based on the examples of other currency notes shown to the Committee on Technologies to Deter Currency Counterfeiting, there appear to be no major technical hurdles to the introduction of clear plastic windows and other ACS-based features. However, the introduction of a window or other ACS will likely affect the manufacturing operations at the BEP.

Since the ACS is a concept of a feature platform rather than any particular technology, it is not possible to provide estimates of cost. However, many of the technologies that could be leveraged for an ACS-type feature are under development for large-scale industrial, medical, and military applications. So it is reasonable to assume that cost-effective manufacturing (including cost minimization) is a major goal of many of these advanced materials-based projects. Therefore, it is possible that at least some ACS-based features will have the same cost as that for the current security thread.

Further Reading

While no specific references are given here, numerous examples of smart materials under development may be found at the National Nanotechnology Institute's Web site at <www.nano.gov>. Accessed February 2007.

CHEMICAL SENSORS

Description

Sensors embedded in banknotes could detect a human-produced or gadget-produced chemical and generate a human-detectable signal. Passive sensors would change their appearance directly, while active sensors would require a power source that could be either self-generated on the banknote or obtained from a battery, for instance at the point-of-sale. Active sensors can trigger visual, audible, or tactile responses (see the section below entitled "e-Substrate"). This feature class can enable features for unassisted use or assisted use with simple devices, as well features for the blind.

The sensed chemical could be an exhalation gas, activated by breathing on the sensor. Expired air has typically 3.6 percent carbon dioxide as compared with 0.03 percent in ambient air, making it a good target. Expired air also has 6.2 percent water given an atmospheric air water content of 0.5 percent.[1]

The sensed chemical could also be one of the constituents of perspiration, activated by touching the sensor. Perspiration is 98 to 99 percent water, but also contains (per 100 ml perspiration): lactic acid (45 mg to 452 mg), chloride (30 mg to 300 mg), sodium (29 mg to 294 mg), and potassium (21 mg to 126 mg) as well as numerous other organic and inorganic compounds in smaller quantities.[2] The sensed quantity could also be acidity or alkalinity. As an example, a lactic acid sensor could be triggered when someone touched the note, then either the sensor's appearance would change or the sensor would supply an electric current to activate a light or sound. Care would be required when designing the sensor to ensure that it is reversible, that is, that the sensor returns to its original state after the trigger chemical is removed.

Alternatively, a gadget similar to the commonly used "starch pen" could be designed to contain a chemical that produces a temporary change in the banknote's appearance or to produce an audible or tactile signal. An example could be a pen filled with vinegar or some other inexpensive, nonhazardous substance that could trigger a reversible response when drawn across the banknote.

[1] For additional information on the human respiratory system, see <http://users.rcn.com/jkimball.ma.ultranet/BiologyPages/P/Pulmonary.html>. Accessed February 2007.

[2] For additional information on latent fingerprint composition, see the information from the Victoria Forensic Science Centre, Victoria Police, Australia, available at <http://www.nifs.com.au/F_S_A/Latent%20fingerprint%20composition.pdf>. Accessed February 2007.

Feature Motivation

This feature platform received a high rating from the committee because of its potential to deter opportunist, petty criminal, and professional criminal counterfeiters, and because of its potential usefulness for the unassisted general public, cashiers, tellers, and the blind, either unassisted or with the use of an inexpensive device. The primary benefit of this feature platform is the difficulty of reproducing or simulating it. Chemical sensors are difficult to reproduce by opportunist counterfeiters and petty criminals because neither the sensors nor the materials required to make them are readily available in the marketplace. Professional criminal counterfeiters would also be deterred by the difficulty of reproducing these sensors well.

The primary limitation of this feature platform is its potential complexity, which might make it expensive to manufacture and limit its robustness over the expected life of the banknote. The sensors must be activated by all people (for example, young, old, healthy, sick) in the full variety of habitable environmental conditions (for example, wide ranges of humidity and temperature).

The effectiveness of chemical sensors for banknote authentication depends entirely on how sensitive and robust the sensors are and on the specific implementation of the human-detectable response. In general, active features (those that change in response to a stimulus) should be easily noticed by the general public and should even generate interest in observing the note.

Potential Implementations

Passive sensors could be formed from chemically activated optical materials sandwiched between porous plastic films; for instance, the material's refractive index could normally render it transparent, but the material would change to opaque upon exposure to lactic acid or carbon dioxide. Or, the material could change color, polarization, or thickness.

Electrical sensors could be formed from electronic elements whose electrical properties would change reversibly when the concentration of a specific chemical was changed. These elements could be transistors, diodes, or passive components. This class of sensor does not produce a detectable signal itself, but rather it provides a trigger to activate a separate, human-sensible device such as a buzzer, light, or raised bump.

Specific examples of these sensors are given below.

Scenario 1a. Passive Sensors: Liquid Crystals

Liquid crystals sandwiched between porous plastic films could be fabricated so that one end of each crystal was attracted to a specific chemical and the other end was repelled. The plastic films could be finely grooved to ensure that the liquid crystals were initially aligned with their long axes oriented parallel to the plane of the banknote. The presence of the target chemical penetrating through the film would rotate the crystals by 90 degrees, rendering them perpendicular to the film. This rotation could be used to produce a number of features: (1) If the plastic films were linearly polarized and formed a transparent window through the banknote, the liquid crystals could be aligned so that they had crossed polarization in their initial state, forming a dark window, and then would lose their polarization in their perpendicular orientation, forming a bright window. (2) The liquid crystals could be chiral and oriented so that they were transparent in one orientation and would create a spectrally pure interference color in the other (switching from transparent to brightly colored). This feature could be used either in reflection or transmission, allowing its use as either a windowed feature or a patch.

Scenario 1b. Passive Sensors: Optical Interference

A material can be sandwiched between porous plastic films in such a way that the films form an interferometer known as a Fabry-Pérot cavity. One film must be highly reflective, the other partially so. Light reflected back through the partially reflective surface will have a characteristic, spectrally pure interference color. Upon exposure to a trigger chemical, either the refractive index of the sandwiched material will be altered or the material will swell slightly. Either mechanism will cause a change in the color of the reflected light.

For another implementation, the sandwiched material chosen can have a refractive index initially the same as that of the films but that changes upon exposure to a trigger chemical. An internal reflection that can be viewed obliquely would exist in the triggered state. The reflection would disappear upon removal of the chemical. This effect might also be used to enable and thwart thin-film interference or to render the material opaque in its high-index state.

Scenario 1c. Passive Sensors: pH-Sensitive Dyes

Material whose color changes with changing pH can be selected for sandwiching between porous plastic films. A person's touch would change the color of a patch of this material.

Scenario 1d. Passive Sensors: Fluorescent Molecules

Material phosphorescence is highly sensitive to local oxygen concentration. In general, the less oxygen, the brighter the phosphorescence. A sensor can be made that has bright phosphorescence under ultraviolet (UV) illumination, but much less bright phosphorescence under the same illumination immediately after exposure to an exhaled breath. This feature requires a gadget, namely, a UV light source.

Many organic compounds are highly fluorescent, and they can be designed so that their fluorescent spectra and/or brightness can change upon exposure to a specific chemical. These changes can be observed under UV illumination in the same way as that described above for oxygen quenching.

Scenario 2a. Electrical Sensors: Transistors

Transistors are commonly used for chemical sensors because of their sensitivity and selectivity. Very briefly, a transistor operates by allowing electrons across a barrier in a controlled fashion. When the transistor is designed to enable specific chemicals to enter the device, the chemicals can alter its resistivity and hence affect the electron flow—that is, the transistor becomes a chemical sensor. These devices are now being made on the microscale, and research is ongoing on nanoscale devices (for example, see Wang et al., 2006). Nanofabricated transistors offer the potential for very low power consumption and highly robust operation on flexible substrates.

Scenario 2b. Electrical Sensors: Diodes

Thin-film Schottky diodes have recently begun to be used as chemical sensors (Gergen et al., 2001). Gas molecules impinge on a thin metal layer and generate electric current. The diodes are built so that they can be chemically selective. Currently, Schottky diodes require electrical heating to function and thus are likely not practical for banknote use. Research continues to reduce the power draw, so they may become viable. Since these diodes are best used as gas sensors, they could be good for detecting exhaled breath.

Scenario 2c. Electrical Sensors: Passive Components

Passive components such as resistors and capacitors can also be used as chemical sensors, but their sensitivity is generally very low.

Materials and Manufacturing Technology Options

Passive sensors are integrated into a banknote by the embedding of an optical material between two porous plastic films. These films can form a window or they can be adhered to one surface of the banknote. (See the preceding section, "Anomalous Currency Space.")

Electrical sensors can be formed by printing different kinds of organic semiconductors in arrays using the multiple nozzles of an ink-jet printer. Or they can be embedded into a banknote by depositing flexible, coated wires.[3] (See the section "e-Substrate," below.)

Simulation Strategies

Claiming that the sensor is broken is probably the best way to "simulate" this feature, for all classes of counterfeiters. Even if the selected sensor has near-perfect reliability, the public may not believe this and may be very willing to accept that all things electronic break.

Otherwise, as chemical sensors become ultrareliable and inexpensive, they may become available commercially and simply be removed from other products for use on banknotes. A target market for the widespread use of chemical sensors is that of food packaging, for the detection of spoilage. Another growing market is biological sensors for homeland security. Care must be taken in selecting an activation chemical for banknote use that is unlikely to be used by these industries.

Key Development Risks and Issues: Phase I

Significant development issues must be addressed for this feature to be viable as a counterfeit deterrent. The most critical issues are the following:

- The selection of the human-produced chemical to detect that covers the range of human variability.
- The long-term durability of the sensor and its power source.
- The cost of the sensing system.
- The accuracy of the sensor over the range of operating conditions and multiple sensing attempts.
- The establishment of the fact, with extremely high confidence, that the sensor will not collect and transmit human pathogens.

[3]For additional information on woven transistors, see <http://www.coe.berkeley.edu/labnotes/0204/lee.html>. Accessed February 2007.

Development Plan: Phase I

The committee determined that the development challenges for chemical sensors were sufficiently high, compared with those for the other feature concepts, that pursuing a development program at this stage would be premature.

Estimate of Production Costs

The production costs cannot reasonably be estimated at this point. The concept of chemical sensors as counterfeit-deterrent features is quite immature at the present time.

Further Reading

Assisi, F.C. 2005. UC Berkeley's Vivek Subramanian Invents Electronic Nose. Available at <http://www.indolink.com/printArticleS.php?id=120105035208>. Accessed February 2007.

Gergen, B., H. Nienhaus, W.H. Weinberg, and E.W. McFarland. 2001. Chemically induced electronic excitations at metal surfaces. Science 294: 2521-2523.

Wang, L., D. Fine, D. Sharma, et al. 2006. Nanoscale organic and polymeric field-effect transistors as chemical sensors. Analytical and Bioanalytical Chemistry 384(2): 310-321.

DIGITALLY ENCRYPTED SUBSTRATE

Description

The concept of the digitally encrypted substrate involves adding small-diameter optical-fiber segments to the substrate. These fiber segments, when illuminated by laser light or narrow spectrum illumination, create a unique signature that can be tagged to the specific note or material. This feature can be used by a robust authenticating machine reader. To employ this feature, optical fibers, or more preferably fiber segments, are placed in the substrate. As the substrate is manufactured, these fiber segments become mixed in the paper batch before the paper is dried. Since the mixing is a random process, the fiber pattern in the substrate will be very unique from note to note.

When the finished substrate is illuminated with light, especially laser light, the fibers will light up as the incident light emanates from the ends of the fibers. The first deterrent example would be for a user to notice the speckles of light from the substrate when it is illuminated. Figure D-1 illustrates the fibers embedded in the substrate. See the section "Further Reading," below, for further illustration of the concept and use of this technique.

When the manufacturing cycle of the banknote was nearly completed, a selected region of the note would be scanned or digitally photographed. This image would then be converted to a secure, two-dimensional bar code that would be printed on the banknote. In order to authenticate the note, a machine reader would compare the image of the selected region to that stored in the encrypted bar code. This method would be extremely difficult to copy, since each note is unique, and re-creating the exact fiber pattern in the substrate would be virtually impossible. The limitation of this approach is the need to scan, photograph, or otherwise capture a picture of the substrate and process the picture.

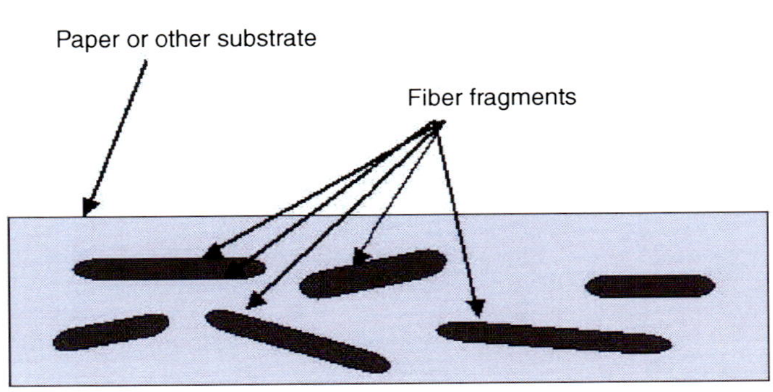

FIGURE D-1 Fiber-infused substrate.

Feature Motivation

This feature has a high rating from the committee owing to the difficulty of exactly duplicating the feature—namely, the fibers in the substrate—and the utility of the highly robust image analysis of the fiber placement. Furthermore, this feature would not be reproducible using electronic printing and scanning techniques and hence would frustrate nearly all counterfeiters. It is expected that only a few very persistent counterfeiters would attempt to generate their own substrate with fibers in it. If they did, they still could not duplicate the exact fiber structure in authentic currency substrate materials.

Potential Implementations

The discussion under "Description," above, provides the scenario of interest. A less secure version of this feature is summarized in Appendix C, in the section "Fiber-Infused Substrate."

Materials and Manufacturing Technology Options

The manufacturing requirements for this feature would involve paper manufacturing and integrating the appropriate-sized fiber fragments into the paper or other substrate material. The fibers could be glass, plastic, or micro- or nanomaterials, with custom design of the properties as required.

It is not expected that adding the fibers during the papermaking process would be a difficult operation, although some tooling and process changes would no doubt be required at the substrate manufacturing plant. Once the substrate was produced, note production at the BEP could proceed as usual until the end, when two additional steps would be needed: pattern scanning of the selected region and printing of the encoded unique image on the banknote.

Simulation Strategies

Simulation of this feature by would-be counterfeiters would be nearly impossible. Furthermore, only the professional criminal or state-sponsored counterfeiter would have any hope of even embedding fibers in the substrate. Since the BEP could also control the fiber materials in the substrate, the counterfeiter would be faced with the difficult task of creating the fibers as well as making the substrate, which would rule out the vast majority of criminals; only state-sponsored counterfeiters would have any plausible chance of attacking this feature. While would-be counterfeiters could attempt to put something on or in the currency to simulate the effects of the fibers, in most cases the counterfeiter may not know what is being evaluated when the banknotes are properly authenticated. Therefore, while simulating the look and feel of real banknotes may be attempted, authentication by an approved device or instrument being used would be an exercise in frustration for most counterfeiters, even those who are state-sponsored. Ultimately, the security of this feature would depend on the strength of the encryption algorithm and the allowable tolerance in the degree of pattern match.

Key Development Risks and Issues: Phase I

The durability of this digitally encrypted substrate feature is unknown, but it would depend on the lengths of fiber embedded in the currency as well as on the continued sharpness of the unique image as the banknote became worn. For instance, if the fibers broke or became disbonded from the substrate, the image could change significantly. No degradation of the fibers themselves is anticipated. Should some of the fibers be conductive—that is, metallic—breakage may not be an issue but the fibers should not be able to stick out of the note by being too long. Combinations of optical- and electronic-fiber characteristics could compound the difficulty faced by the would-be counterfeiter. It is also possible that semiconducting fibers could be used in conjunction with the e-substrate methods (see the section entitled "e-Substrate," below).

There should be no aesthetic issues with this feature. Until illuminated, the note would look and feel identical to one without the feature. Even when illuminated, the feature should not detract from the note's appearance. The two-dimensional bar code would not be very large—perhaps 1 mm × 1 mm—and should be printed in an inconspicuous location such as the margin of the note. Thus, this feature would be aesthetically neutral, conforming to the look and feel of current notes.

Similarly, there should be no social acceptability issues surrounding this feature, with the caveat that the public must be convinced that scanning the note and authenticating it would not reveal any privacy information.

The key technical challenges include the following:

- The selection of the optical fiber that possesses the necessary characteristics for incorporation into the substrate.
- The selection of the appropriate three-dimensional random pattern and encryption of the pattern in such a way that an authenticator can be printed directly on the banknote.
- The ability of the fibers to survive the high-pressure intaglio process without unacceptable breakage.
- The durability of the pattern itself, and of the printed encryption of the pattern.

- The implementation of a method to read the pattern quickly and to compare the results with the printed authenticator.
- The development of a hierarchy of effects that provide increasing security with increasing authentication capabilities.
- The cost of implementing the feature.

Development Plan: Phase I

The maturity level of this technology is relatively low. The science behind its use and verification is known, but this technique has not been implemented in high-volume, low-cost applications such as that envisioned here.

Key milestones would include the following:

- Place fiber fragments in paper substrates to see the applicability of the technique and any operational or manufacturing difficulties that might arise.
- Scan and visually observe the fragments according to the referenced techniques to see how difficult the practical implementation and verification of this feature might be.
- Evaluate different encryption algorithms using criteria defined by the development team, and select the best.
- Determine the best bar code and location for printing the encrypted image.
- Initiate a discussion with vendors regarding the prototype scanners and associated processing software.

Similar methods were used in placing fiber-embedded placards on missiles for treaty verification. If a placard was removed, the fibers would be broken in places and the imaging pattern from the placard would be altered, making it obvious that the placard was no longer authentic.

Estimate of Production Costs

The optical fibers would be embedded by the substrate manufacturer. This should be a low-cost operation. At least two additional steps would be required at the BEP for the "signing" of the note based on the digital data acquired from the fiber substrate. This might alter the printing procedures or serial number generation, but the operation could be highly automated. It would require new capital equipment at the BEP.

The cost of adding the fibers and signing the note should be approximately the same as that of the current security thread.

Further Reading

Chen, Y., M.K. Mihcak, and D. Kirovski. 2005. Certifying authenticity via fiber-infused paper. ACM SIGecom Exchanges 5(3): 29-37.

DeJean, G., and D. Kirovski. 2006. Certifying authenticity using RF waves. Paper presented at IST Mobile Summit, Myconos.

National Research Council. 1993. Counterfeit Deterrent Features for the Next-Generation Currency Design. Washington, D.C.: National Academy Press, pp. 74-75, 117-120.

ENGINEERED COTTON FIBERS

Description

The cotton fiber is a complex biological structure engineered by both natural selection and intensive plant breeding. Cotton is the premier natural fiber for textile applications. It is a biological composite of cellulose, small quantities of hemicellulose, pectins, and proteins that provides excellent wearability and aesthetics. Cotton fiber is normally hollow, although many fibers collapse after drying or later during processing.

Engineered cotton fibers are not a feature but rather a set of tools that may be employed to generate a wide array of potential features. Throughout this description, reference to cotton fibers is understood to mean fibers treated in a manner similar to that for the materials currently used in FRNs. These cotton fibers have already undergone extensive processing, first to form clothing and then to form the composite paper substrate used by the BEP. The durability and other materials characteristics of the cotton fiber used in FRN production and the test modes for determining them are well known. The current substrate material is effectively a cotton-linen composite. Using recombinant deoxyribonucleic acid (rDNA)-based or conventionally bred cotton fibers would not change that.

The various ways of engineering a cotton fiber include adding new materials to the fiber lumen, modifying the cellulosic material that forms 90 percent of the fiber itself, modifying the proteins associated with the fiber, or a combination of these methods.

For example, a second biopolymer can be synthesized within the fiber lumen without affecting fiber wall integrity. Researchers have been able to use genetic engineering to fill the hollow center (lumen) with a natural thermoplastic polyester compound, poly-D-(–)-3-hydroxybutyrate (PHB) for synthesis in fiber (Maliyakal and Keller, 1996). The new cotton fibers exhibited measurable changes in thermal properties that suggested enhanced insulation characteristics. The engineered fibers conducted less heat, cooled down slower, and took up more heat than conventional cotton fibers.

Feature Motivation

Engineered cotton fibers offer the opportunity for the FRN to retain that unique, highly distinctive feel that is frequently cited as a major attribute of the U.S. dollar, while also promising application as counterfeit-deterrent features. Custom-engineered cotton varieties could be developed for any number of applications, from physical identification by the end user to embedded, cryptic forensic features. Among the advantages of engineered cotton fibers are the following:

- The impossibility of counterfeiting or simulating resulting features by all but state-sponsored counterfeiters.
- The unique spectroscopic signatures that can be achieved by modifying protein R-groups.
- The possibility of "filling" fiber with other compounds, which opens up significant possibilities to specialize the fiber further.
- The significant amount of current research regarding enhancing properties of the fiber.

A wide range of naturally occurring materials show physicochemical properties that could be useful as features. The feature could involve an addition of new materials to the fiber lumen, modification of the cellulosic material that forms 90 percent of the fiber itself, modification of the proteins associated with the fiber, or some combination. These include the following:

- Proteins that contain iron and other electromagnetic and/or paramagnetic metals (for example, hemoglobin).
- Proteins that contain optical and/or electro-optical properties (for example, bacteriorhodopsin).
- Proteins that form fibers to add strength, toughness, or stiffness to the fiber (for example, actin).
- The large list of natural compounds with useful properties that could be "loaded" into the lumen. When one adds in the possibility of creating synthetic genes that encode synthetic proteins or other materials for lumen loading, the possibilities become extremely large.

Potential Implementations

This subsection describes some examples of potential applications for custom-engineered cotton fibers. One possible implementation is the creation of a series of cotton varieties whose fibers have surface-associated proteins genetically engineered to display novel epitopes (molecular structures) sufficiently different from other cotton varieties so that rapid, powerful immunoassays could be used for identification purposes. Immunodetection tools, in turn, vary both in sensitivity and ease of use. Simple tools could be developed for the end user—for example, an "immunoassay pen" similar to the starch-detecting pen. This type of instrument would give a quick (seconds) colorimetric reaction that would be highly accurate. However, a positive reaction might be simulated by, for example, grinding up a single FRN and dusting multiple counterfeit bills. In such a case, the immunoreaction would obviously be weakened, so successful detection would depend on the rigor with which the "immunoassay pen" was used. Engineering of multiple novel epitopes into a single cotton variety or the blending of multiple varieties would create an extremely complex immunoreactive profile for which a complementary mixture of monoclonal antibodies could be generated for law enforcement and other rigorous detection applications. In this latter case, an immunological profile would be generated with sufficient redundancy to make counterfeiting or simulation totally impossible. The sensitivity of immunoreactions is such that nondestructive microsampling could be employed even in crucial forensic applications.

Another implementation would take advantage of the structure of a normal cotton fiber that is similar to a hollow tube. rDNA technology is currently being used to create fibers whose hollow center (lumen) is filled with various types of chemical compounds. The properties of any specific material used to fill the lumen will, in turn, affect the properties of the fiber and, ultimately, the FRN. Once again, a range of features is ultimately possible. One example would be that loading the lumen with, for example, a small iron-containing protein such as hemoglobin would ab initio introduce certain measurable electromagnetic and paramagnetic properties. This property might be manipulated further, for example, by passing the FRN through an electromagnet powerful enough to align the spins of all the molecules in a region of the bill or even of the entire FRN. This type of feature would be highly amenable to machine reading. Incorporation of virtually any unique molecule into the lumen will create an equally unique molecular fingerprint for forensic applications. With current instrumentation, only microsampling would be required to see this molecular profile via, for example, gas chromatograph-mass spectrometry.

Proteins associated with the surface of the cotton fiber could also be modified via rDNA technology to provide electromagnetic, electro-optical, or other unique signatures. A simple example would involve changing the numbers and types of R-groups on these proteins to specifically change certain fluorescence and/or absorbance properties. These modified proteins would, in turn, change the "signature" of the FRN with respect to its interaction with certain forms of electromagnetic radiation. Some of these signature changes could be relatively simple and amenable to hand detection technology: for example, tryptophan fluorescence. Other changes in signature could be more complex: for example, changes in circular dichroism spectra based on changes in amount of helicity in the proteins present on the fiber surface.

Regarding nonspecific physicochemical characteristics of the cotton fiber, any large plant-breeding program will produce and retain a large number of individuals (genotypes) whose target phenotype (in this case the cotton fiber) is undesirable because this individual expresses some other desirable phenotype—for example, insect resistance or drought tolerance. Therefore, it is highly probable that a substantial number of genotypes already exist that produce unusual cotton fibers. As a result, it should be possible to rapidly create new, stable varieties of commercial cotton that produce unusual or abnormal fiber phenotypes (measurable physical traits such as fiber length, strength, roughness, lumen size, and so on). Because the cotton fibers produced by certain engineered genotypes were undesirable for commercial applications, the resulting cotton varieties were either put aside by breeders and molecular geneticists or only used in early crosses during the development of commercial varieties. Some of these traits may be of interest for FRN production—for example, fibers with increased surface roughness. In almost all such cases these traits will be characterized by the breeding programs only in terms of classical genetics, meaning that the genes involved have not been identified or even mapped at the molecular level. As a result, a situation emerges where key enabling technology—the gene or genes that encode the trait—remains unknown. The BEP could control information about, or even patent, the genes for a fiber trait useful for currency production as they are cloned and characterized.

As an example, suppose that the cotton plant could be engineered to produce fibers with bacteriorhodopsin. Bacteriorhodopsin is found in the intensely purple cell membrane of a bacterium called *Halobacterium salinarium*, which grows in salt marshes. Illuminating the protein triggers a photochemical reaction cycle, which transports protons along a channel spanning the cell membrane. The membrane's purple color comes from a bacteriorhodopsin component called retinal, which is strongly bound to an amino acid inside the membrane channel. Unbound retinal in solution is pale yellow. Alternating laser light of two different wavelengths on the protein molecule can switch it back and forth between its purple and yellow forms. That behavior has prompted research on the use of bacteriorhodopsin as the light-sensitive element in artificial retinas and as memory or processing units.[4]

Fibers containing genetically engineered bacteriorhodopsin could be throughout the paper or in specific areas, thereby creating a color-changing technology that would be virtually impossible to counterfeit or simulate, especially if multiple forms of bacteriorhodopsin were used so that two or more colors changed simultaneously. Work on the bacteriorhodopsin system was the subject of a chapter in a recent report from the National Research Council (NRC, 2001).

The rating of any particular genetically engineered trait will depend entirely on the materials engineered into the fiber and the properties these materials display. Inherent in the genetic engineering strategy is the degree of difficulty required to isolate, synthesize, and clone the genes to produce cotton in such a manner that the desired materials are expressed and properly targeted to the fiber.

The ability of any particular genetically engineered (or even naturally occurring) trait to deter counterfeiting will depend entirely on (1) the properties resulting from the incorporation of the materials into the cotton and (2) any properties resulting from the processing of the engineered cotton into the paper for

[4]See <http://www.sciencenews.org/pages/sn_arc97/3_8_97/fob2.htm>. Accessed February 2007.

the FRN. As an example, consider a bacteriorhodopsin-based trait that will deter counterfeiting by exhibiting complex dynamic behavior that is only displayed by these classes of materials (see the following subsection). The bacteriorhodopsin would be engineered so that in natural or artificial light, one color (or spectrum of colors) is displayed, whereas covering the FRN (or placing it in darkness) will terminate the color-generating process and cause the bacteriorhodopsin to revert to its baseline color. It is assumed that the ability to synthesize and fabricate such complex, dynamic materials will *not* be possible for any but the most sophisticated counterfeiter—that is, state-sponsored. Obviously, if there are other synthetic dyes that are similar in behavior to bacteriorhodopsin, then the feature could be simulated. It is assumed that such dyes would be equally complex to synthesize and/or obtain, so that only the most well-funded sophisticated counterfeiter could simulate this trait.

One scenario involves genes encoding engineered bacteriorhodopsin (rBR) being incorporated into a specific cotton line and maintained under appropriate security. The plant line has been engineered so that the rBR is expressed and incorporated to high levels in the lumen of the cotton fiber. While the light-blocking properties of such placement would have to be determined, it is probable that if the FRN contained a large proportion of this type of fiber, the FRN would display (at the least) a plainly visible background color that changed when the bill went from light to dark. Use of rBR genes engineered for different color-shift properties might allow two or even more color-shifting background tones to be obtained.

A second scenario involves genes encoding engineered rBR being incorporated into a specific cotton line and maintained under appropriate security. The plant line has been engineered so that the rBR is expressed and incorporated on the surface of the cotton fiber at levels sufficient to produce a highly visible color shift but low enough to allow the fiber to maintain its general physicochemical properties (as manifested by length, strength, toughness, and so on). It is probable that if the FRN contained a large proportion of this type of fiber, the FRN would display a plainly visible surface color that changed when the bill went from light to dark. Use of rBR genes engineered for different color-shift properties might allow two or even more color-shifting colors to be obtained.

There are a number of possibilities that range from simple to highly complex. On the simple side, modification of cotton fiber proteins to contain additional fluorescent biological R groups (for example, tryptophan) creates an opportunity to incorporate complex fluorescence patterns directly into the paper itself. This includes taking advantage of fluorescence shifts of up to 20 nm by modifying the environment of the engineered fibers during paper formation. On the complex end, the creation of cotton plants that produce fibers containing nonbiological components—metal nanoparticles and so on—offers the opportunity to create patterns with unique or uniquely complex electromagnetic or optical properties.

Materials and Manufacturing Technology Options

The features discussed above are in the long-range category, so it is not possible at this time to describe specific manufacturing strategies for currency applications. However, some strategies are already in development for other applications. For example, protein-linked quantum dots are currently in use as fluorescent and/or color-generating tags for a number of applications, with more in development.[5]

Simulation Strategies

The rBR-based feature described above should be impossible to simulate by all but state-sponsored counterfeiters. To simulate this feature, the counterfeiter would have to (1) know the specific gene construct and what it encoded or be able to deduce it from the behavior of the FRN, and (2) reproduce the effect created in a bona fide FRN via the use of rDNA-based cotton fiber.

Key Development Risks and Issues: Phase I

It is reasonable to assume that at least some of the useful phenotypes that may be created in a cotton fiber by rDNA technology will be as durable as those displayed by the natural fibers. For example, materials incorporated into the lumen of the cotton fiber should, in general, remain intact as long as the fiber itself remains intact. Likewise, novel epitopes created by engineering fiber-associated proteins should, in general, display the same wear properties as those of the fiber-associated proteins in natural cotton. The final molecular structure of these epitopes will be formed as fully hydrated biomolecules are extracted, processed, and ultimately dried down to form paper. Just as normal FRN paper wears and smoothes with age, FRNs containing fibers produced by engineered cotton will be expected to show normal wear characteristics, except where modification involves exotic or extreme changes in the biochemistry and/or supramolecular structure of the fiber. The stability of these more extreme products would need to be closely examined, and appropriate cost-benefit analyses would need to be conducted with respect to any trade-off between anticounterfeit value and decreased durability.

In terms of reasonable analogies, there are now numerous examples of the use of rDNA to modify or enhance desirable plant phenotypes. Because cotton is a major source of both food and fiber in the United States, genetic engineering of this plant is now routine. Most of the commercial work has focused on relatively simple single-gene traits such as herbicide or pest resistance (for example, Monsanto's

[5]See <http://nano.cancer.gov/news_center/nanotech_news_2007-01-8c.asp>. Accessed February 2007.

incorporation of a gene that allows the plant to survive the application of the herbicide glyphosate). Given the existence of an efficient transformation system (which does exist in cotton), the speed with which new traits may be incorporated into this plant will be a direct function of the complexity of the molecular genetics underlying that trait and whether any or all of the genes involved have been cloned and characterized. However, there is certainly a great deal of technology both in the public sector (U.S. Department of Agriculture [USDA]) and private sector that could be used to leverage and accelerate progress.

In terms of aesthetics, the use of genetically engineered plants has been generally accepted in the United States, especially for nonfood applications, so there is little probability that any type of panic would be created if GM cotton fibers were incorporated into U.S. currency. However, this possibility cannot be entirely discounted. The use of rDNA technology might even impart a high-technology gloss to the FRN that would be appealing to some sectors of the public. In addition, the use of genetically engineered cotton has the potential to enhance the security of the FRN while allowing the BEP to retain most of the traditional look and feel of the "American greenback." Retention of the traditional craft involved in creating U.S. currency and the reliability that has accrued to the dollar's "brand identification" are, in and of themselves, highly aesthetically desirable. Obviously, any change in fiber characteristics will need to be evaluated for its impact on the look and feel of the FRN. Traits specifically designed to affect the look and feel (for example, increased surface roughness) will need to be evaluated in even more detail for aesthetic impact.

The environmental consequences of growing genetically engineered crop plants has been under constant evaluation for more than 25 years. Appropriate federal and state regulations have been developed that, in all probability, will cover the type of limited acreage required by the BEP for FRN paper substrate. Security considerations—that is, the need to protect the cloned genes, the cotton germplasm (seeds), the plants in the field, and so on—would provide additional de facto insurance against any unintended release of novel plant genes into the environment.

The key technical challenges will depend entirely on the trait(s) selected by the BEP for genetic engineering. As discussed above, the manufacturing issues directly associated with genetic engineering of a new phenotype into cotton have been largely addressed. Likewise, the parameters involved in determining the stability of a plant phenotype during the creation of a "true breeding" commercial plant variety are extremely well known and applied routinely. In other words, as with all major crop plants in the developed world, the infrastructure for cotton production and processing already exists as a mature manufacturing industry with well-established Technology Qualification Application procedures, and so on.

The outcome of this study will allow a rational decision-making process with respect to what (if any) type of program should be initiated in this area.

Development Plan: Phase I

Activity during Phase I must address the key technical challenges mentioned above. There are many opportunities for features within this technology area, so Phase I activities can determine which potential features are of greatest interest for counterfeit deterrence. Feasibility experiments can be conducted, through a combination of laboratory-based work and modeling and simulation analysis, to select the most promising future directions for currency applications.

A thorough review of current related work would be the first priority. For instance, it is likely that multiple candidate features exist as phenotypic traits that have been produced as by-products of long-term, ongoing cotton breeding programs throughout the nation and the world. This is equivalent to data mining of existing inventories (libraries) of chemical compounds by pharmaceutical companies when a new potential application is identified or new screening technology developed.

Because cotton has been the subject of intensive, long-term conventional breeding programs, it is possible, even probable, that conventional genetic methods have already produced fibers with phenotypes that could translate into useful features for currency. These traits could be characterized and the fibers used without ever engineering the genes involved, so long as the germplasm and plants remained secure. Engineering genes offers the further possibility of introducing completely novel biomolecular structures that fall outside of the metabolic capabilities of the cotton genome (for example, genes that encode a system that forms fibers where the lumen is filled with paramagnetic proteins).

There are development programs underway in many countries in both the public and private sectors with the goal of using biotechnology to modify various aspects of cotton fiber structure and development. The United States is currently the leader. The cotton fiber itself is the target of much of this work. In addition, there is an effort to create synthetic genes that encode proteins with unusual traits and/or cellular production of, for example, protein-nanoparticle composite materials (biomolecular-materials composites), some of which would undoubtedly be useful as FRN features.

The BEP could work directly with the appropriate departments within the USDA (for example, the Agricultural Research Service [ARS]) and/or with appropriate private-sector consultants to survey the current state of knowledge and capability with respect to the modification of cotton fibers (including lumen loading). Also, the nanomedicine initiative at the National Institutes of Health (NIH) involves a significant effort in the creation of biomolecular-materials composites.

The technical requirements for producing engineered cotton are well known. A realistic scenario is that the novel GM cotton varieties would be produced in an existing facility using existing equipment. By analogy, the phenotype encoded by the new genes would most likely be verified and parameterized by existing research facilities. One possible path would be a research partnership with the USDA or its ARS. Alternatively, the work could be done by private industry or even in a university research facility, so long as appropriate security precautions were maintained.

Use of engineered or genetically unique cotton is a long-term project even if varieties currently exist that produce fibers with novel phenotypes that could provide the basis for novel features. Since there are so many possibilities of new features, multiple feature concepts can be pursued. Therefore, there could be a base level of activity that is continually operating in Phase I to further develop new engineered cotton fiber options. Then, specific ideas that have been determined to have attractive counterfeit-deterrence benefits could be spun out of the base program into a distinct, defined Phase I, II, and III project.

The time and effort to develop the new line depends on a number of factors, which include the following: the complexity of the phenotype selected, whether the genes encoding that phenotype have already been engineered, the extent to which the products encoded by these genes have already been characterized, and so on. For instance, if cotton genotypes useful for counterfeit deterrence already exist, the feasibility of incorporating those new fibers into the currency substrate could be demonstrated within 3 years. If they do not exist, the time line could easily be doubled to 6 years.

Estimate of Production Cost

The major technical challenges to the incorporation of engineered or genetically novel cotton fibers into FRNs involve the substrate producer. Also, modification of the physicochemical properties of the fiber could change the substrate enough to impact BEP operations. This compatibility issue should be addressed during Phase II of the development process as the BEP updates the substrate material specification. Once the new cotton line is developed, the incremental production cost should be extremely low, less than the cost of the current security thread. It is highly unlikely that the BEP would need to invest in any new capital equipment in order to generate new features based on engineered cotton fibers.

References and Further Reading

Maliyakal, E.J., and G. Keller. 1996. Applied biological sciences metabolic pathway engineering in cotton: Biosynthesis of polyhydroxybutyrate in fiber cells. Proceedings of the National Academy of Sciences 93(November): 12768-12773.

National Research Council. 2001. Opportunities in Biotechnology for Future Army Applications. Washington, D.C.: National Academy Press. Available at <http://www.nap.edu/catalog/10142.html>. Accessed February 2007.

e-SUBSTRATE

This feature can be subdivided into passive and active electronic substrates, here called "e-substrates."

Passive e-Substrate

Description

Passive electronic substrates—that is passive e-substrates—refer to classes of features that are enabled by techniques and materials currently used in large-area electronics (for example, in liquid-crystal televisions) or low-cost electronics (for example, in radio-frequency identification tags [RFIDs]), as well as those that are being explored for newer devices such as flexible displays. This type of feature consists of passive structures (i.e., those without electronic functionality) fabricated with techniques and materials similar to those used for these electronic systems, in the form of security threads or patches integrated with the paper substrate. The collections of materials—dielectrics, conductors, semiconductors—that can be patterned using these approaches and the resolution and registration accuracy that can be achieved in single and multilayer configurations provide visible as well as invisible attributes that have value for currency applications owing to the high levels of difficulty in replication or simulation. In this version of the e-substrate feature, the patterns do not offer any form of electronic or optoelectronic operation. The subsection below on "Active e-Substrate" examines the functional possibility.

Feature Motivation

Currency is now produced, almost exclusively, by conceptually old printing techniques (for example, offset, intaglio, and so on) that have historically dominated the conventional printed paper industry. The nature of emerging threats and the powerful technologies that are becoming available to the counterfeiter suggest that it might be useful to examine whether this basic approach to currency production will likely continue to provide a path to secure currency indefinitely into the future. In particular, one can argue that intrinsic limits in resolution, registration, pattern layouts, and inks associated with printing methods may make it necessary to consider radically different manufacturing concepts. Through the passive e-substrate feature platform, a fundamentally new approach is examined here that adapts, for the production of currency or currency features that offer no electronic functionality, tools and fabrication facilities principally designed for electronics used in large-area applications, such as liquid-crystal televisions and computer monitors, or low-cost devices, such as RFID tags. The latter devices use circuits that consist of patterns of dielectrics (for example, silicon dioxide and silicon nitride), metals (for example, aluminum), and semiconductors (for example, amorphous silicon) formed on large-area glass substrates with layouts designed to switch individual pixels in these displays. The basic approaches are, however, sufficiently adaptable that they can be implemented with a wide range of other substrates (for example, plastics), materials (for example, polymers), and designs that could be useful for passive features in currency in the form of patterns or images with overt or covert security purposes.

The technologies for producing this kind of electronics, sometimes referred to as macroelectronics (Reuss et al., 2005, 2006), are advancing rapidly, as measured in production capacities, costs per unit area, yields, and electrical performance. Although the main application drivers are flat-panel displays, emerging applications in structural health-monitoring equipment, x-ray imagers, sensors, and other systems also exist and are increasing in significance. As a result, in addition to commercially established fabrication approaches, newer techniques and materials are under development at many small and large companies worldwide, with goals of further increasing and lowering the areas and costs, respectively. Taken together, these recent developments suggest that it may be possible to consider existing or emerging macroelectronic-like processing techniques as next-generation methods for producing passive images or patterns for currency features or, ultimately, the entire currency note itself.

This approach has many attractive features. First, the fabrication techniques offer levels of resolution (~500 nm) and pattern alignment (~500 nm) that are better, by at least 10 times, than anything that is likely achievable with established printing techniques. Second, the wide range of materials that can be processed into patterns, in single-layer formats or complex multilayer stacks, lie outside of the possibilities associated with printed inks. Third, the approach provides a scalable manufacturing platform with user-definable levels of complexity in the materials and pattern layouts, as well as new active forms of currency features, such as those that require active electronics, light emitters, solar cells, sensors, and other systems with diverse functionality, as described in the subsection below, "Active e-Substrate." Fourth, the high capital costs and technical sophistication of the fabrication facilities would prevent any counterfeiter, other than those associated with well-financed, state-sponsored organizations, from reproducing (or simulating, depending on the design) currency or currency features produced in this fashion. Fifth, the current technology status and future trends in costs, capabilities, and production capacities indicate that this approach to currency production is, for certain implementations such as security strip features, feasible now, with costs that are expected to continue to decrease in the future.

In macroelectronics, the area coverage defines the metric by which progress is measured, rather than critical feature sizes or integration densities as in conven-

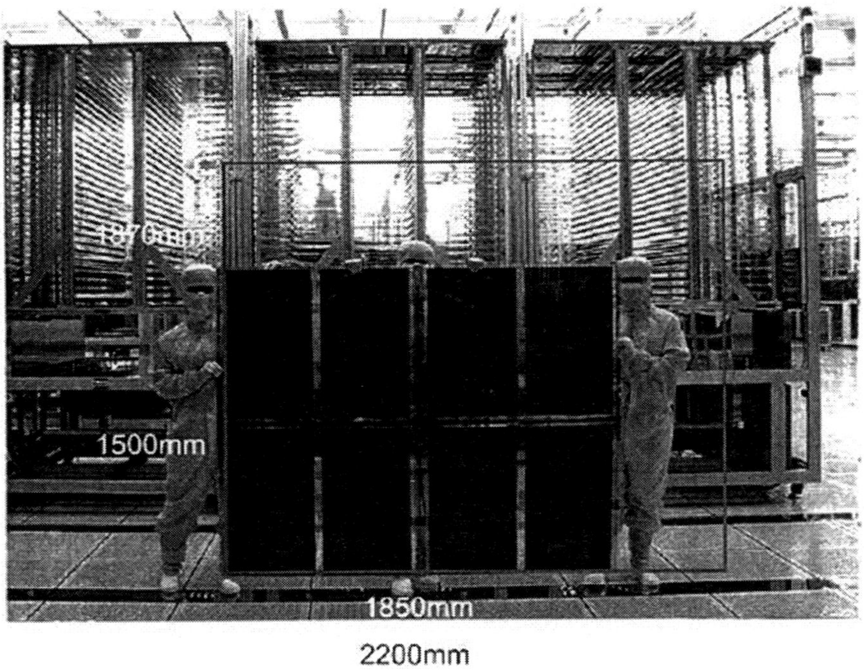

FIGURE D-2 Large-area circuits in the Gen 7 format (see discussion in text). SOURCE: Reuss et al. (2005). © 2005 IEEE.

tional wafer-based integrated circuits. Some of the first commercial macroelectronic systems, which appeared in 1987, used substrates (glass) with sizes of 270 mm × 200 mm, known as Gen 0 glass, and were implemented in 8.4 inch liquid-crystal displays. Currently, Gen 7 glass is in production (1,870 mm × 2,200 mm), and some Gen 8 (2,200 mm × 2,500 mm) fabrication facilities became operational in 2006.[6] Figure D-2 provides an image of a completed circuit on Gen 7 glass. These large-area systems are patterned using photolithographic processes with step-and-repeat stages capable of stitching together multiple images. Whereas the integrated-circuit industry uses mainly single-crystal silicon in wafer form, the semiconductor of choice for existing macroelectronic systems is sputtered thin-film amorphous silicon (a-Si). Deposition, etching, and patterning of this material, and other vacuum-deposited materials needed for the circuits (for example, gate insulators,

[6]See <http://www.corning.com/jp/en/media-center/press_releases/2006/2006050102.aspx> and <http://displayblog.wordpress.com/2006/08/01/sharp-g8-starts-august-two-months-early/>. Accessed February 2007.

metal interconnects and electronics, encapulants, and passivation coatings), can be accomplished over very large size scales. Figure D-3 shows an optical micrograph of one of the many millions of thin-film transistors that exist on substrates like the one shown in Figure D-2. A device of this type typically requires the sequential deposition and patterning of between four and six layers of material.

The proposed concept is that these, or other classes of material structures, could be used as passive, visible or invisible, patterns or multilayer-structured elements on a currency note. Alternatively, they could be used in conjunction with conventional pick-and-place methods to integrate separately processed substrate pieces using strategies similar to those used for RFID tags. This approach provides another pathway to the passive e-substrate feature. Very recently, the costs of both routes to such systems have reached levels that can be contemplated for currency or currency features. The most realistic possibility, as argued subsequently, involves the use of such approaches for the fabrication of new, passive features for conventional printed paper notes, where these features, in the form of narrow strips, are cut from macroelectronic substrates and integrated with the paper using approaches currently employed for digital optically variable devices and security strips.

Potential Implementations

In the simplest implementation, the patterns formed using the materials and patterning techniques of macroelectronics provide passive features that have a distinct appearance (for overt operation) or that incorporate hidden information that is invisible to the unaided eye (for covert operation). The design flexibility, in

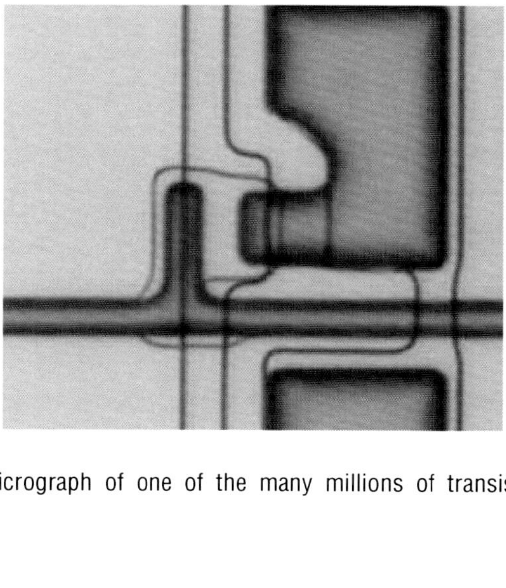

FIGURE D-3 Optical micrograph of one of the many millions of transistors in a liquid-crystal display.

terms of pattern layouts and materials choices, are large. The methods are already well developed for existing applications in large-area and low-cost electronics, as outlined above. Implementation in currency would involve, then, the specification of designs that provide the desired levels of overt and covert functionality, with attributes that maximize the difficulty of simulation. Creating these designs represents the first phase of work in implementing this class of feature.

Materials and Manufacturing Technology Options

The cost structure of the passive e-substrate features is critically important to their implementation in currency. The following discussion is limited to existing, commercial-scale fabrication approaches based on thin-film processing for the passive features. (Some of the same arguments apply to technology options for the electronics and/or interconnects, the active features described in the subsection below, "Active e-Substrate.") The level of detail on new methods for fabricating electronics that use printable materials and techniques similar to printing is relatively low, thereby preventing a quantitative analysis of their possible use for these applications. These methods will, however, become increasingly important as future technology options as their state of development improves.

The photolithographic process for patterning is central to the fabrication of all forms of electronics that have achieved widespread use. This process involves passing ultraviolet light through a mask to expose, in a patterned geometry, a thin layer of photosensitive polymer known as a photoresist. Washing away the exposed (or unexposed, depending on the chemistry of the photoresist) regions creates a pattern of photoresist that can then serve as a sacrificial mask for spatially directing the deposition or etching of other materials to produce the final structures. As implemented in large-area electronics, this process offers extremely high resolution (~500 nm) and layer-to-layer registration capabilities (~500 nm), when compared with conventional printing techniques. The methods used to dice the large-area substrates into smaller pieces for device integration are also well developed. Although most available data apply to patterns formed on the sorts of glass substrates that are used in displays, the basic processes are applicable to a range of other substrate types, including flexible plastics or paper substrates with suitable planarizing and protective coatings. These types of materials should, if sufficiently planar and smooth, allow resolution that is comparable with that achievable on glass. Mounting such substrates onto glass carriers for ease of handling would allow them to be processed with minimal modifications to the tooling. The registration capabilities, however, would likely be degraded somewhat owing to dimensional instabilities (for example, from thermal expansion and contraction and other mechanisms) in the substrates. Also, current systems are most fully developed for patterning only on one side of the substrate. New handling equipment would have to be developed for processing both sides. Given the intrinsically high resolution and low levels of distortion associated with the patterning techniques, it is reasonable to expect that suitable mechanical designs and optical registration schemes (that is, alignment marks in transparent regions of the substrate) could yield good front-to-back registration.

The materials used in addressing circuits for liquid-crystal displays—amorphous silicon, silicon dioxide, silicon nitride, metals (for example, aluminum, copper, and so on), and others—are typically deposited by physical or chemical vapor deposition or electroplating (electrolytic or electroless). Deposition through openings in photoresist masks followed by removal of the resist leaves patterns of these materials. Patterning can also be accomplished by using these masks, to prevent etching with wet chemical baths or dry plasmas in reactive ion etching tools. Versions of these and other deposition and etching methods can be used—for example, those used to coat plastics with metals to improve hermeticity in food packaging or to provide optical substrates for data storage (for example, CDs and DVDs) or reflective diffractive optical elements that are at present used in currency. The photoresist masking procedures are applicable to broad classes of other materials that might be considered for currency applications. In addition, many materials themselves can be designed to be photodefinable, for direct patterning by photolithography, without the etching or deposition steps. For example, conducting polymers, semiconducting small molecules, electroluminescent polymers, and dye doped polymers have all been successfully patterned in this manner. Multilayer stacks of patterned materials can be produced by repeating any of these processes. Detailed technical requirements (for example, for resolution, registration, defect density, yield, and so on) for passive e-substrate features would likely be much different, and more relaxed, than those for functional circuits, depending on the designs. As a result, the range of materials and substrates that could be used is likely to be broad.

Simulation Strategies

The difficulty of simulating a passive e-substrate would be a strong function of the design of the feature. The manufacturing approaches for these features provide, as described above, a level of flexibility in materials choices and of patterning capabilities that would be difficult or impossible to reproduce by all classes of counterfeiter (with the possible exception of the state-sponsored class), owing to the high capital cost and technical sophistication of the manufacturing systems. Effective implementation in covert features would be straightforward. Use in overt or visible features would require optimizing designs to frustrate various simulation strategies.

Key Development Risks and Issues: Phase I

For the production of currency or currency features, the adaptation of processing systems similar to those used for electronics represents a radical departure from conventional printing-based manufacturing. Such a radical change has associated risks. The approaches described here are, however, inherently technically feasible for passive features. The durability of a passive e-substrate strip would be comparable to, or better than, that of a diffractive optically variable device (DOVD) or conventional security strip.

Development Plan: Phase I

The implementation of the passive e-substrate feature would require adapting for this application the existing processing approaches outlined above. A key part of the development is the definition of suitable feature designs and materials that frustrate simulation strategies. Also, the modes of manufacturing would need to be considered. The preferred approach would be to create a dedicated manufacturing facility based on a design similar to a flat-panel-display fabrication line but different in key ways. Other issues, such as durability, should be examined. The durability of the passive e-substrate feature is estimated by the committee to be comparable with, if not better than, current holographic features.

Estimate of Implemented Production Cost

The costs and capacities for manufacturing of large-area electronics have recently reached levels at which they can begin to be considered for currency or currency-feature fabrication. The latest facilities, some of which first became operational in 2006, use glass substrates that measure 2,200 mm × 2,500 mm (Gen 8). Initial configurations of these manufacturing plants have capacities of ~15,000 Gen 8 substrates per month, with projections of about 30,000 to 50,000 substrates per month at full capacity.[7] The most promising integration pathway uses passive patterns formed on thin plastic substrates (mounted on the glass or a rigid substrate for processing) that are then integrated into conventional printed paper currency through the use of strips woven into or bonded to the paper. If these strips are 5 mm wide and 5 cm long, then ~20,000 of them can be cut from a single Gen 8 substrate. This number corresponds to a monthly production capacity of ~1 billion strips per month, from a single fabrication facility in its initial configuration, and about 2 billion to 4 billion strips per month at full capacity. These numbers are

[7]See <http://displayblog.wordpress.com/2006/08/01/sharp-g8-starts-august-two-months-early/>. Accessed February 2007.

conservative estimates, since many currency implementations (for example, passive images or structures) would require only one or two patterned layers instead of the four to six that are used for circuits. Some of this advantage disappears, however, if both sides of the substrate must be patterned.

In addition to the production capacities, the costs are critically important. Figure D-4 presents the cost per square meter of fully formed pixel switching circuits for color liquid-crystal display (LCD) modules, as a function of year, based on estimates associated with a Gen 7 fabrication line. For 2006, these estimates correspond to costs of \$0.08 per strip if it is assumed that the additional costs of processing on plastic sheets and of cutting and packaging the elements for currency are no more (or less) than those for making the displays. These costs represent upper bounds for the passive features, because they require the following; (1) only one or two patterned layers, in the simplest embodiments, which is three or four times fewer than those needed for the circuits; (2) only modest yields and levels of registration, compared with those needed for active circuits; and (3) classes of plastic substrates that have less demanding requirements than the ultraflat glass plates used for circuits. These differences might, the committee speculates, reduce the costs by a factor of about 5 times—for example, 2 times for (1) and another of 2.5 times for (2) and (3)—bringing the estimates to ~\$0.015 per strip. Finally, adopting such a manufacturing approach for currency or currency-feature fabrication would happen, at the earliest, about 5 years from now. Projections as shown in Figure D-4 suggest that costs in 2010 could be about 30 percent lower than they are today, leading to costs of \$0.01 per strip and about \$0.45 per note. The cost of any

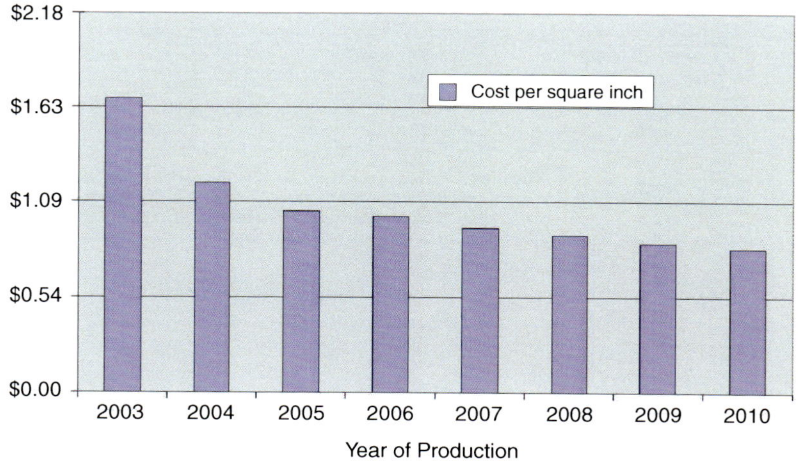

FIGURE D-4 Estimated cost per square inch of circuits in the Gen 7 format as a function of production year, including projections, based on model-generated data. SOURCE: iSuppli, Inc., Flat Panel Display Manufacturing Cost Model: TFT-LCD (February 2005).

specific implementation would scale accordingly, by area. Successful efforts in the development of newer and potentially lower cost approaches to forming large-area electronics could lead to further reductions in cost. These numbers indicate cost feasibility for the strip embodiment of the passive e-substrate feature.

References and Further Reading

Reuss, R., B.R. Chalamala, A. Moussessian, M.G. Kane, A. Kumar, D.C. Zhang, J.A. Rogers, M. Hatalis, D. Temple, G. Moddel, B.J. Eliasson, M.J. Estes, J. Kunze, E.S. Handy, E.S. Harmon, D.B. Salzman, J.M. Woodall, A.A. Alam, J.Y. Murthy, S.C. Jacobsen, M. Olivier, D. Markus, P.M. Campbell, and E. Snow. 2005. Macroelectronics: Perspectives on technology and applications. Proc. IEEE 93(7): 1239-1256.

Ruess, R.H., D.G. Hopper, and J.-G. Park (eds). 2006. Macroelectronics. MRS Bulletin 31(6).

Active e-Substrate

Description

Active electronic substrates—that is active e-substrates—refer to classes of features that, like the passive e-substrate features, are enabled by techniques and materials emerging from developments in large-area and/or low-cost electronics.

Unlike the passive features, however, the active e-substrate provides functional electronic or optoelectronic devices (for example, light emitters, chemical sensors, or actively programmable surface topography) that interact actively with the cash handler (for example, by emitting light or sound or by sensing chemicals and emitting a signal of some sort) or with a machine reader (for example, by responding to a radio frequency or to microwave radiation). This active functionality provides, together with the structures themselves, extremely high levels of security. This subsection provides some conceptual designs for active structures, including ideas on the power generation required and the production of human perceptible responses. Some information on manufacturing options is also included. The committee stresses that the technical challenges in meeting the durability and cost requirements of such features for currency applications are extreme. The purpose of this subsection is only to introduce some notional concepts.

Feature Motivation

The active e-substrate feature platform represents a fundamentally new approach to currency feature design in which parts of the note are able to interact actively with the general public or with specialized machine readers. The specialized manufacturing approaches, which are similar to those described for the passive e-substrate feature, together with the active functionality lead to features that offer extremely high levels of security. Duplication or simulation would be possible only by large, state-sponsored counterfeiters. This class of feature, while extremely challenging technically, can begin to be considered owing to the many recent advances in large-area and low-cost electronics, as described in the subsection above, "Passive e-Substrate."

Potential Implementations

Potential implementations of the active e-substrate feature have considerations that overlap and extend beyond those associated with the passive embodiments. The active functionality could provide significant benefits, if successfully implemented. Achieving acceptable cost structures and device durability and providing adequate power supply are formidable technical obstacles. Potential scenarios for the three primary e-substrate components for active functionality (power generation, circuitry, and human-perceptible response) are described below. Specific features would use one method from each component type—for example, a piezoelectric film embedded within the banknote could generate a current that is delivered to an organic light-emitting diode (LED) by means of inorganic flexible electronics that produces a "twinkling eye" in the note's portrait when the banknote is squeezed between a cash handler's thumb and finger.

Scenario 1a. Power-Generating Devices: Piezoelectric

- *Source of energy:* Pressing (squeezing) or shaking the banknote.
- *Principle of operation:* Certain materials generate electrical power when compressed. The amount of power is proportional to the piezoelectric coefficient and thickness of the material, and to the applied force. Piezoelectric films, disks, and rods are activated by compression (pressing or squeezing the banknote), and piezoelectric cantilevered beams are activated by shaking. Common piezoelectric materials include quartz and lead zirconate titanate, and polymers are now being developed as piezoelectric elements for microelectromechanical systems (MEMS) devices.[8]
- *Power-generation capability:* It has been reported that 200 microwatts of power have been generated from a microfabricated piezoelectric cantilevered device vibrated by resting on the outside case of an operational

[8]For more information on the integration of piezoelectric materials into MEMS devices, see <http://www.npl.co.uk/materials/functional/pdf/mems_metrology.pdf>. Accessed February 2007.

microwave oven (Roundy, 2003). Brief power spikes of 120 milliwatts for 0.1 millisecond have been generated from 15-micron-diameter fibers embedded in a 5.85-millimeter slab by dropping a 34-gram ball bearing from a height of 10 centimeters (Mohammadi et al., 2003).

- *Durability/manufacturing considerations:* Monolithic piezoelectric devices are commercially available as rods or cylinders and can be embedded into the banknote in the same way as the current security threads, except now lead wires are required. The lead wires and piezo rods would be embedded during the same manufacturing step. Alignment tolerances are a function of the cross-sectional area of the piezo rod. Good electrical contact is required, but the orientation of the wire and piezo rod is not critical. Alternatively, the piezoelectric rod and wires could be surface printed on the banknote using standard ink-jet technology. A microfabricated piezoelectric cantilever could be similarly embedded into the banknote, but in this case the wire must be precisely aligned with the piezo device. This device could also be surface printed but requires three-dimensional control using, perhaps, rapid prototyping technology.
- *Potential research topics:* Research will be needed to investigate the production of materials with higher piezoelectric coefficients.

Scenario 1b. Power-Generating Devices: Reverse Peltier (Thermoelectric) Device

- *Source of energy:* One possible energy source is the temperature gradient between the user (either directly or through an input device) and the ambient environment—possibly by means of holding a banknote between the fingers, breathing on the banknote, or placing it in a warm or cold location.
- *Principle of operation:* The reverse Peltier effect, known as the Seeback effect, discovered in 1822, produces a voltage when the device is exposed to a temperature gradient, with the voltage produced proportional to the temperature difference. A traditional Peltier device converts electrical current into temperature gradient. As an electrical current is passed through a junction made up of two materials with different conductivities, heat is either produced or absorbed at the junction. A second junction, which completes the electrical circuit, displays the opposite heating or cooling. This thermoelectric effect, which is much more dramatic in semiconductors than in metals, is currently seeing use in a range of heating and cooling applications—for example, in small coolers that can be run off an automobile's 12 volt power outlet.

- *Power-generation capability:* The power-generating capability of a thermoelectric device is not entirely clear. A number of issues affect the potential for small-size applications such as in currency applications. The power density of a thermoelectric device is inversely proportional to the thickness (distance between the hot and cold surfaces). A number of companies are currently using nanotechnology, including quantum dots, to produce micron-scale devices.[9] Such devices are currently being used for small-scale cooling, but would conversely be capable of scavenging energy. One research effort currently has the following goal for a high-density nanoengineered thermoelectric material: generate 100 microwatts of power at 3 V with 1°C temperature change on a 6 cm^2 surface.[10]
- *Durability/manufacturing considerations:* Manufacturing and durability considerations similar to those for piezoelectric devices would apply for thermoelectric devices integrated into currency, with the exception that it is unlikely that ink-jet technology would be capable of directly printing thermoelectric devices onto currency.
- *Potential research topics:* Extensive research is ongoing to improve thermoelectric figures of merit and efficiencies. Research should focus on the ability of current technologies to produce devices capable of powering low-power outputs such as LED for currency applications.

Scenario 1c. Power-Generating Devices: Solar

- *Source of energy:* External light source (sunlight, flame, or electric light).
- *Principle of operation:* Solar cells convert energy from light into electrical current either directly through the photovoltaic effect or indirectly through the generation of heat. Most solar cells use the photovoltaic effect and generate a voltage across a p-n junction by using the energy from incident photons to release trapped electrons, allowing them to flow and generate a current through a circuit. The photons can come from any radiant source, provided they have sufficient energy (brightness).
- *Power-generation capability:* Typical conversion efficiency of commercial single-crystal solar cells is about 16 percent, with a theoretical maximum of about 28 percent and laboratory-demonstrated efficiencies of about 24 percent (Moslehi, 2006). Silicon crystalline solar cell wafers about 0.3 mm

[9]See <http://www.nanocoolers.com> and <http://www.evidenttech.com>. Accessed February 2007.

[10]For additional information, see Biophan Technologies at <http://www.biophan.com/2004meeting/riedlinger_TEBio_presentation.ppt>. Accessed February 2007.

thick, with a diameter of 10 cm to 15 cm, generate approximately 35 mA of current per cm^2 area at a voltage of 550 mV at full illumination (Lenardic, 2001).

- *Durability/manufacturing considerations:* Emerging thin-film technologies based on the deposition of very thin films of photovoltaic materials on plastics could offer much-lower-cost alternatives to silicon-based cells. A large group of materials is being considered for thin-film solar cells, including cadmium telluride, cadmium sulphide, and copper indium diselenide. A compound semiconductor-based approach uses single-crystal gallium arsenide and its alloys, such as gallium-indium phosphide. Other technologies being pursued include low-cost plastic solar cells (Moslehi, 2006). Recently, titanium dioxide thin films have been developed for potential photovoltaic cell construction. These transparent films are particularly interesting because they can also serve double duty as windows (Parry-Hill et al., 2006).
- *Potential research topics:* Extensive research is ongoing to improve photovoltaic conversion efficiency and durability.

Scenario 1d. Power-Generating Devices: Capacitive

- *Source of energy:* Pressing (squeezing) or shaking the banknote.
- *Principle of operation:* An electrical capacitor stores electronic charge for release when required. As envisioned for banknote power generation, the capacitor can amplify a small initial voltage by harvesting mechanical energy obtained by squeezing or shaking the banknote. A capacitor consists of two charged plates separated and insulated from each other by a dielectric material. The amount of capacitance, *C*, is inversely proportional to the thickness of the gap, and can thus be dynamically changed by changing the gap via squeezing or shaking. Note that the gap can be either in the plane of the banknote or along its thickness. The voltage across the capacitor is inversely proportional to the capacitance, so the voltage is proportional to the gap thickness.
- *Power-generation capability:* An optimized, in-plane gap, microfabricated variable capacitor has been demonstrated to generate 100 microwatts per cubic centimeter (Roundy, 2003).
- *Durability/manufacturing considerations:* Variable capacitors can be manufactured using traditional MEMS fabrication methods. The durability issues are similar to those for thermoelectric devices.
- *Potential research topics:* Research is needed to improve the durability of a microfabricated device for banknote use.

Scenario 1e. Power-Generating Devices: Inductive

- *Source of energy:* Rubbing two ends of the banknote together, or passing the banknote across an external magnet, or shaking the banknote.
- *Principle of operation:* Current is generated by moving a coiled wire past a magnet. Both the magnet and wire can be embedded in or on the surface of the banknote, and a current can be generated by moving one across the other. The wire can be on one side of the note and the magnet on the other, and current would be generated rubbing the ends across each other. Alternatively, the magnet could be an external, point-of-sale device and current would be generated by swiping the note across the magnet. Finally, the coil and magnet can be integrated into a microfabricated device embedded within the note, and current would be generated by shaking the note. The amount of current is proportional to the number of winds in the coil, the coil diameter, and the strength of the magnetic field.
- *Power-generation capability:* For a fixed magnetic strength, the voltage generated (V) is proportional to the number of coil winds (N), the strength of the magnet (B), and the rate of change with time of the area common to the coil and the magnet (A): $V = -N\,B(\Delta A/\Delta \mathrm{t})$. This voltage is likely to be very small for a banknote-embedded device, because the limited size of the magnet will limit both the magnetic strength and available area. The number of coils can be made large using microfabrication techniques.
- *Durability/manufacturing considerations:* The "wire" can be created using microfabrication techniques to provide a large number of "coils" in a very small area. Containing the wire in a very small area improves durability by reducing the local radius of curvature for even severe crumples. Alternatively, very flexible wire can be used and distributed over a relatively large portion of the banknote. "Rubber metal,"[11] gold-wound threads, and other technologies being developed for electronic textiles[12] are applicable. The Department of Defense sponsors research in this area that might be relevant.
- *Potential research topics:* Research is needed to improve the durability of a microfabricated device for banknote use.

[11]For more information on rubber metal, see <http://www.sciencentral.com/articles/view.php3?type=article&article_id=218392354>. Accessed February 2007.

[12]For more information on electronic textiles, see <http://www.research.ibm.com/journal/sj/393/part3/post.html>. Accessed February 2007.

Scenario 1f. Power-Generating Devices: High Power Density Batteries

- *Source of energy:* Chemical or nuclear.
- *Principle of operation:* Nuclear microgenerators spontaneously emit high-energy electronics that can be used to generate paired electrons and holes to produce current across a diode. Nickel-63 is a safe choice because its relatively low energy beta particles are easily absorbed by a 25-micrometer layer of plastic; they are also absorbed by the thin dead-skin layer covering our bodies (Lal and Blanchard, 2004).
- *Power-generation capability:* One nuclear battery has been shown to generate about 3 nanowatts of power from 0.1 millicuries of nickel-63 (Lal and Blanchard, 2004). Additional power has been generated from devices combining radioactive-decay with piezoelectric amplification (Lal and Blanchard, 2004).
- *Durability/manufacturing considerations:* Nuclear batteries are currently expensive: 1 millicurie of nickel-63 costs about $25, but the cost of tritium for a single microbattery might be as low as a few cents because it is a by-product from some nuclear reactors (Lal and Blanchard, 2004).
- *Potential research topics:* Research is needed to reduce the cost and increase the power density of nuclear microbatteries.

Scenario 2a. Electronics: Printed Electronics

One strategy for forming the electronics component of an active e-substrate feature involves the deposition and patterning of the circuit layers directly on the substrate (for example, plastic for a security strip). The development of semiconductors that can be deposited over large areas and the patterning techniques for building circuits out of them are central to this approach. The most-established technology is based on amorphous silicon (a-Si), deposited by physical vapor deposition, for the semiconductor and photolithographic tools for the patterning. This process, which involves vacuum deposition steps and high temperatures, is extremely well developed for glass substrates (for example, for active matrix liquid-crystal displays), and it has also been demonstrated, in slightly modified forms, on flexible plastic substrates. The main disadvantage is that the performance is relatively low, limited by the poor mobilities in amorphous silicon (~1 cm^2/Vs). The performance can be enhanced considerably (mobilities >100 cm^2/Vs) by the use of pulsed laser annealing techniques to convert the amorphous silicon into large, oriented-grain polycrystalline silicon. Although this method must be applied with care to avoid thermal damage (for example, thermal buffer layers are often used), polycrystalline silicon-based circuits have been achieved on plastic substrates (that is, polyimide) and flexible steel foils in this manner. Other semiconductor materials include organic polymers and oligomers, as well as organic-inorganic hybrid materials, which exhibit mobilities of 0.1 to 1 cm^2/Vs. Transistors formed with solution-deposited inorganics (that is, nanoparticles or nanowires) have similar performance: effective device mobilities (as determined empirically using parallel plate capacitances defined by the physical dimensions of the channel) of ~1 cm^2/Vs for the nanoparticles and ~2 cm^2/Vs for the nanowires. The intrinsic mobilities of the nanowires can have values approaching those of wafer scale sources of material. Films of solution-grown inorganics, such as CdSe, represent another possibility; the mobilities in this case are ~1 cm^2/Vs and higher, depending on the materials and deposition techniques. Finally, carbon nanotubes, in the form of thin films of aligned arrays of tubes or random networks of them, can also be used for these systems. Devices built with such semiconductors can exhibit, in some cases, close to the extremely high intrinsic mobilities of the tubes (several thousand cm^2/Vs). Table D-1 summarizes a few of the options. Figure D-5 shows some flexible circuits formed using inorganic and organic semiconductors. The judgment of the committee is that the technology that is already well developed for active matrix liquid-crystal televisions represents the most promising potential path to circuits for the e-substrate feature. The newer approaches that use printable semiconductors and other materials are currently not sufficiently well developed for serious consideration for currency applications.

Progress in these areas will, of course, potentially change this situation. Figures D-6 and D-7 show circuits formed by printing.

TABLE D-1 Summary of Selected Semiconductor Materials That Have Been Implemented in Flexible Electronic Systems

Semiconductor Material	Mobility (cm^2/Vs)	Processing Technique	Technology Demonstrations	Technology Status
Small molecule organics	<5	Evaporation	Large areas, circuits	Small-scale fabrication
Polymers	<1	Solution printing	Large areas, circuits	Small-scale fabrication
Hybrid organic-inorganic	~1	Solution printing	Small areas, devices	Inactive
Amorphous silicon (a-Si)	~1	Evaporation	Large areas, circuits	Commercial
Laser annealed a-Si	~100	High temperature	Small areas, circuits	Research
Solution CdSe, SnS_2	~10	Solution printing	Small areas, devices	Research
Grown nanopart	~1	Solution printing	Small areas, devices	Research
Grown nanowires	~1	Solution assembly	Small areas, devices	Research
Nanotube networks	~50	Solution printing or transfer printing	Large areas, circuits	Research
Nanotube arrays	~200	Transfer printing	Small areas, devices	Research
Microstructured silicon	~500	Transfer printing	Large areas, circuits	Research

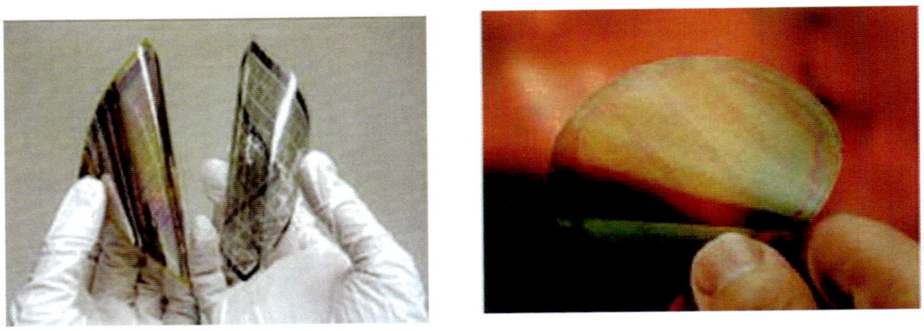

FIGURE D-5 Flexible electronics based on (left) inorganic (Service, 2006) and (right) organic semiconductor materials (Drury et al., 1998). SOURCES: (Left) Courtesy of I. Chun Cheng, Princeton University. (Right) Reused with permission from Drury et al. (1998), ©1998, American Institute of Physics.

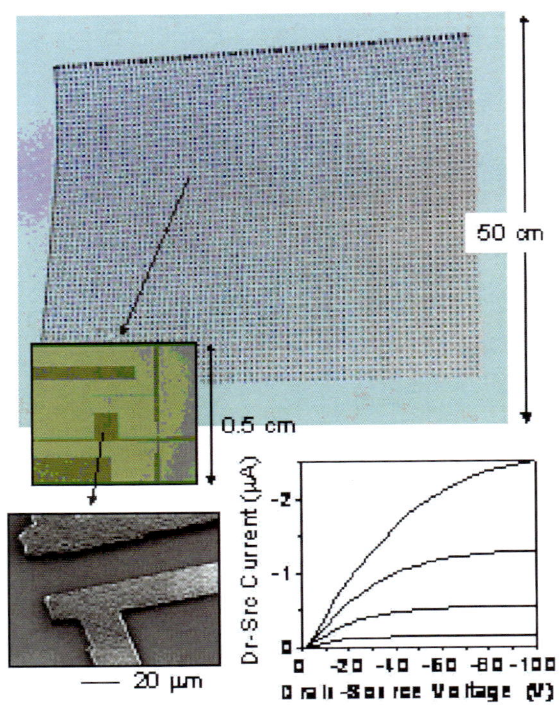

FIGURE D-6 Images and performance characteristics of a large-area printed circuit, formed with a laser thermal printing technique, and an organic small-molecule semiconductor. SOURCE: Reused with permission from Blanchet et al. (2003), ©2003, American Institute of Physics.

FIGURE D-7 Flexible, paperlike displays that integrate electrophoretic inks with printed flexible electronic backplane circuits. The images show (top left) an initial demonstration and (bottom left) a printed backplane circuit (Rogers et al., 2001); and (right) an example of the current state of the art (<http://www.plasticlogic.com>).

Scenario 2b. Electronics: Silicon Blocks

There are several methods for dicing wafers into very small pieces (for example, 100 × 100 μm, or in some cases much smaller—2 μm × 100 nm) and then integrating them on plastic substrates to form flexible, low-cost devices. The most well-established technique uses robotic "pick-and-place" tools to move these elements from the source wafer to the target substrate. Many existing forms of RFID tags, smart cards, and related applications are formed in this manner. Newer approaches rely on solution suspensions of these elements and guided self-assembly techniques[13] to achieve this pick-and-place outcome without the need for precision robotics. At least one company is manufacturing RFID tags in this manner, although the cost per device is relatively high. Recent research publications describe yet another approach, in which soft elastomeric stamps pick up many ultrathin

[13]For more information, see <http://www.alientechnology.com>. Accessed February 2007.

elements (sometimes referred to microstructured silicon, μs-Si) chemically etched from a source wafer in a parallel fashion and transfer them, in a single-step printing process, to a target substrate. High-performance devices and circuits have been demonstrated using this approach, on a range of substrates including rubber and thin plastic sheets. For all three of these approaches, the semiconductor is usually single-crystal silicon, although many other materials are possible. Electrical interconnects are typically patterned on the device substrate to interface the elements with one another or with other subsystems such as the power sources described in the previous subsection.

Scenario 3a. Human-Perceptible Response: Light-Emitting Elements

With the "twinkling eyes" feature—a light shines from the banknote (the eyes "twinkle")—the light could be a single-colored point, or an array with a variety of colors. One approach would be to use an organic light-emitting-diode (OLED) powered by a MEMS piezoelectric device activated by shaking. OLEDs consist of electroluminescent organic layers sandwiched between a transparent anode and a substrate-mounted cathode. Excluding the substrate, which may be <1 mm thick, the thickness of the rest of the layers is typically 20 nm to 50 nm total. OLEDs are being developed for video displays, and some demonstration devices have been achieved on flexible surfaces. Power requirements are given assuming large pixelated arrays and are a function of brightness. State of the art seems to be 250 mW for movie-theater-equivalent brightness for 852 pixels × 3 pixels × 600 pixels. Others quote "low current at 2-10 volts." A single pixel might require only ~250/(852 × 3 × 600) = ~0.2 microwatts. OLEDs can emit anywhere in the visible spectrum, but reds and greens are more efficient than blues. A major challenge with OLEDs in existing devices is their limited lifetime and reliability and the temporal drift in their properties. A promising alternative path uses ultrasmall inorganic LEDs. Such devices can be formed with sufficiently small dimensions for integration onto flexible substrates. They have the advantage of avoiding many of the challenges associated with OLEDs. This inorganic approach might represent the most promising embodiment of light-emitting elements.

Scenario 3b. Human-Perceptible Response: Visible, Nonemissive Elements

Liquid-crystal devices, electrochromic materials, and electrophoretic cells represent some of the non-emissive visible elements that could be considered. These systems have the advantage of requiring much less power than that needed by the types of emissive elements described above. Several demonstrators of these types of systems, on flexible substrates, exist and appear to be moving toward commercialization for handheld electronic devices such as electronic books, personal digital assistants, and cellular telephones.

Materials and Manufacturing Technology Options

The cost structure of the active e-substrate features is critically important to their implementation in currency. The analysis provided in the "Passive e-Substrate" subsection (above) provides one possible framework for estimating the costs associated with the circuit component of an active e-substrate feature. Costs of the active features could be expected to be ~10 times higher than the simple passive structures, when fabricated with thin-film processing techniques. An attractive alternative manufacturing approach for the active features could exploit the integration of small-scale chips of processed single-crystal silicon, similar to those elements used in RFID tags, as described in the previous subsection. For the foreseeable future, such systems likely offer the most promising means to meet the demanding requirements on reliability, performance, and cost. In this approach, most or all of the device processing and materials sets are borrowed directly from the well-established silicon integrated-circuit industry. The transfer printing or pick-and-place methods then integrate these fully formed devices or circuit blocks onto a thin strip on which other printing techniques form the necessary electrical interconnects. These printing methods can range from well-developed screen printing techniques that are used for printed circuit boards, to newer techniques that use laser thermal transfer (see Figure D-6).

These chips can not only provide electrical functionality, but they can also be integrated with sensors, LEDs, or other elements. As with the circuits, there is a range of possibilities. For example, newer classes of small-molecule or polymer organic electroluminescent materials or conventional inorganic micro-LEDs could be used for the "twinkling eyes" feature. The most robust and lowest-cost systems, especially for applications in currency, where the number of active elements might be much smaller than those in a commercial display device, might be achieved with the most-established technology, that is, inorganic micro-LEDs in this case. These devices, like the silicon circuit elements, can be integrated using assembly or transfer printing techniques.

Simulation Strategies

The active e-substrate feature has the potential to provide overt features that would be very difficult to simulate. Indeed, this technology would be such a radical departure from the current production methods that it would decouple advances in digital printing technology from enhanced counterfeiting tools.

Key Development Risks and Issues: Phase I

The development risks associated with the active e-substrate feature platform are significant. A long-term development program is required.

Development Plan: Phase I

There are many opportunities for features within this technology area. Phase I activities should collect sufficient information and data to determine which potential features are of greatest interest for counterfeit deterrence.

Key milestones for Phase I would include the following:

- Define power requirements for a class of human-perceptible responses—for example, design an OLED having sufficiently small size and durability properties; then determine the current and voltage requirements to drive it.
- Define packaging specifications for the power-generation device and electric conductors to withstand expected durability tests.
- Design the prototype system, including power generation, conductors, and packaging, specifying candidate materials and processes.
- Analytically demonstrate the required power generation.

Estimate of Implemented Production Cost

The costs for an active e-substrate feature are difficult to estimate. Based on the detailed costing analysis for the passive e-substrate feature, the committee estimates a minimum of $0.10 per strip.

References and Further Reading

Alien Technology Corporate Web site. 2005. Available at <http://www.alientechnology.com>. Accessed February 2007.

Blanchet, G., Y.-L. Loo, J.A. Rogers, F. Gao, and C. Fincher. 2003. Large area dry printing of organic transistors and circuits. Applied Physics Letters 82(3): 463-465.

Drury, C.J., C.M. Mutsaers, C.M. Hart, M. Matters, and D.M. de Leeuw. 1998. Low-cost all-polymer integrated circuits. Applied Physics Letters 73(1): 108-110.

Lal, S.A., and J. Blanchard. 2004. The daintiest dynamos. IEEE Spectrum, September: 36-41.

Lenardic, D. 2001. Solar cells. Available at <http://www.pvresources.com/en/solarcells.php>. Accessed February 2007.

Loricchio, D.F. 1992. Key force and typing performance. Pp. 281-282 in Proceedings of the Human Factors Society 36th Annual Meeting. Santa Monica, Calif.: Human Factors Society.

Mohammadi, F., A. Khan, and R.B. Cass. 2003. Power generation from piezoelectric lead zirconate titanate fiber composites. Pittsburgh, Pa.: Materials Research Society. Materials Research Society Symposium Proceedings, Vol. 736: D5.5.1-D5.5.6.

Moslehi, B. 2006. Harvesting the Sun: Renewable power generation from photovoltaic solar cells. Micro Magazine, June. Available at <http://www.micromagazine.com/archive/06/04/reality.html>. Accessed March 2007.

Parry-Hill, M.J., R.T. Sutter, and M.W. Davidson. 2006. Solar Cell Operation. National High Magnetic Field Laboratory, Florida State University. Available at <http://micro.magnet.fsu.edu/primer/java/solarcell/>. Accessed February 2007.

Plastic Logic Corporate Web site. 2000. Available at <http://www.plasticlogic.com>. Accessed March 2007.

Rogers, J.A., Z. Bao, K. Baldwin, A. Dodabalapur, B. Crone, V.R. Raju, V. Kuck, H. Katz, K. Amundson, J. Ewing, and P. Drzaic. 2001. Paper-like electronic displays: Large area, rubber stamped plastic sheets of electronics and electrophoretic inks. Proceedings of the National Academy of Sciences USA 98(9): 4835-4840.

Roundy, S. 2003. Energy Scavenging for Wireless Sensor Nodes with a Focus on Vibration to Electricity Conversion. Ph.D. dissertation, Department of Electrical Engineering and Computer Sciences, University of California, Berkeley, May. Available at <http://engnet.anu.edu.au/DEpeople/Shad.Roundy/EnergyScavenging.pdf>. Accessed February 2007.

Service, R. 2006. Inorganic electronics begin to flex their muscle. Science 312: 1593-1594.

NiTi SHAPE MEMORY AND SUPERELASTIC RESPONSIVE MATERIALS

Description

Shape memory materials, most commonly based on the intermetallic alloy NiTi (often referred to as nitinol), offer a number of phenomena that could be used to produce active responses useful for developing human-detectable features for currency security. These phenomena are based on the reversible austenite-to-martensite phase transformation in these alloy systems. Three phenomena based on the thermoelasic martensite have potential for currency security features:

- *Shape memory effect* (SME)—in which deformed structures are recovered with heating.
- *Transformation superelasticity* (TSE)—in which structures can be deformed to large strains but recover their shape when the stress is released.
- *Two-way shape memory effect* (TWSME)—in which a reversible shape change is induced by changes in temperature.

A number of currency features based on this set of technology can be envisioned:

- *Superelastic features:* These could be wires or thin-foil-based patterns (dots, eagles, numbers, buildings, and so on) that could be easily deformed but would spring back to shape. This technology is the same as that used in some eyeglass frames.
- *Shape memory features:* Similar patterned features can be deformed at relatively low temperature, then recovered at higher temperature, such as under the heating from a human finger, incandescent light, or low-temperature heat source. Applied to currency, this approach may be more problematic, owing to the relatively broad ambient temperature range in which currency must be able to work (for example, −20°C to +40°C).
- *Two-way shape memory effect:* Patterned features will change shape when touched, moving in one direction when the human touch increases the temperature and the other direction when the human touch decreases the temperature. However, this effect may be difficult to achieve effectively within the temperature range available using purely human inputs; it may require external heating and cooling.

Feature Motivation

Such features would be quite effective at allowing the general public to identify genuine currency. The features would require the user to input a stress and identify a response, so some education would be required. However, the response should be robust and easy to identify. Such features would also be very useful for the visually impaired.

The shape memory/superelastic nature of NiTi thermoelastic alloys is well documented and understood. Such alloys are used in a range of industrial, medical, and consumer applications. However, their cost is typically rather high. Considerable industrial and academic research is involved with developing these materials for cost-effective, higher-volume applications. While the base metals in these alloys are relatively inexpensive in the base forms in the amounts needed for currency applications, processing can be quite expensive. Thus, the key to these alloys finding application in currency is to be able to produce relatively thin foils and/or films in high volumes at a reasonable cost.

The active nature of features based on thermoelastic martensite features would be a strong deterrent to currency counterfeiting for all but the most sophisticated counterfeiters sponsored by national governments. Primitive, hobbyist, and petty criminal counterfeiters would be hard-pressed to simulate the active response combined with the metallic nature of the features. The professional criminal counterfeiter would have difficulty duplicating the chemistry (stoichiometry) and processing to achieve a suitable response. State-sponsored counterfeiters would need considerable time to duplicate the processing and stoichiometry control.

Potential Implementations

As noted above, thermal elastic martensite alloys would allow verification of a note by means of user input and/or user-detected output. For superelastic alloys, the user would fold or bend or deform a foil or wire and observe, either by feel or vision, the feature recovering its shape upon unloading. Shape memory features would require the observation of a shape recovery (either tactile or visual detection). Two-way shape memory would require either heating or cooling to induce a user-detected shape change.

Scenario 1

A specific example of a superelastic feature would be a wire or security strip integrated into the substrate. The user could fold the feature over. With release of the load, the strip would snap back into its original shape. Similarly, a foil embossed with a recognizable symbol could be deformed but would retain its shape with unloading.

Scenario 2

A shape memory alloy feature could entail a metallic security strip or foil that would show permanent deformation with use. This deformation might come from either the specific user or through general circulation. To verify the authenticity of the note, the user would heat the note (possibly by means of user body heat or with an inexpensive external source). The feature would recover its original, undeformed shape.

Scenario 3

A two-way shape memory feature could entail a foil with an embossed feature or pattern of bumps. This pattern would appear and disappear with changes in temperature. These temperature changes could be user-induced (for example, heating through touch) or via an external device (such as a lamp for heating, a ice cube for cooling).

Materials and Manufacturing Technology Options

Thermal elastic marteniste alloys based on NiTi are readily available from commercial sources. However, implementation of these alloys into currency requires overcoming significant processing challenges. Once produced, it is anticipated that the features would be relatively easy to integrate into currency. Currently, security threads are integrated into U.S. currency, and a number of international currencies have displayed metallic strips in the past (for example, higher-denomination British notes have included both silver and gold strips). Thus, it is anticipated that NiTi metallic strips could be readily integrated using current technology. Likewise, because many currencies currently include holographic foils, one would expect to be able to integrate NiTi foils readily into notes.

Simulation Strategies

Primitive and opportunist counterfeiters would not be able to duplicate and would have a very difficult time simulating the effects of thermoelastic alloy features. Most simulations would most likely be based on using elastic polymers to simulate the superelastic effect. However, the general response of such features would be significantly different in terms of the mechanical response. Furthermore, the metallic nature of NiTi foils and strips would be lost when using polymers.

Shape memory features could be simulated using shape memory polymers (for example, those used in shrink-wrap and shrink fittings). Again, these would not be metallic in nature. In both cases, these counterfeiters would be challenged to produce the undeformed/recovered features accurately if these were embossed.

Petty criminals would be challenged in attempting to counterfeit thermoelastic alloy features. While a range of shape memory and superelastic materials are available commercially, integrating these into notes in a realistic manner with accurate responses would be difficult. More likely, such criminals would attempt to raise the denomination of notes. However, with the specific shapes and topologies of features on each note (for example, embossing of the denomination), along with strategic placement among the notes, raising notes could be made less effective.

Professional criminals would be very challenged by the processing and alloy control necessary to obtain good duplication of the features and responses. State-sponsored counterfeiters could, with time, reverse-engineer the chemistry necessary to obtain the proper temperature, and response behavior. More time would be required to develop the specific processing that would be required to employ a thermoelastic feature in currency.

Key Development Risks and Issues: Phase I

Manufacturing NiTi to the desired shape (strip, embossed foil) requires careful processing with relatively precise temperature control. Furthermore, the transformation temperatures of these alloys are a strong function of Ni-Ti stoichiometry, tertiary alloys content, and alloy purity. Processing bulk alloys to the thin foils required for currency applications requires precise thermal mechanical processing and high-strength rolling mills. Direct production of thin foils using sputtering and other approaches is currently being carried out in the laboratory but will require technological advances for production scale-up to provide the proper chemistry control in conjunction with the large volume necessary for full-rate currency production.

NiTi thermoelastic alloys can recover approximately 7 to 8 percent elastic deformation in the martensitic phase. Within this strain range these alloys can survive millions of deformation cycles. The key to durability in currency applications will be in avoiding deformation beyond this critical strain in the folding and crunching of the note. However, it should be noted that the elastic strain (ε_E) at the top or bottom of a beam (foil) is proportional to the thickness (t) of the member and inversely proportional to the bending radius (ρ)—that is, $\varepsilon_E = t/2\rho$. Thus, for the thin foils anticipated in currency applications, fairly small bending radii should be sustainable; for example, a foil 20 microns thick will not reach a permanent deformation strain of 8 percent until it is bent to a bending radius of 0.125 mm. Furthermore, it is anticipated that some of the deformation from folding and bending will be accommodated in the substrate. Thus, disbanding of the NiTi from the substrate also needs to be evaluated.

NiTi features would have to be attached to and integrated with a note in a manner similar to holographic foils or woven security strips. From an aesthetic point of view, NiTi features would impact the note in the same manner as these features.

NiTi features should not raise significant social acceptability concerns, as they are not electronic and contain components found in coins (Ni) and consumer products (both Ni and Ti). NiTi alloys are currently accepted by the public in eyewear.

Development Plan: Phase I

The basic technology of superelastic and shape memory alloys based on thermoelastic martensite is well developed and understood. A range of industrial and consumer products are currently manufactured with these alloys. The critical activities to be addressed in Phase I include the following:

- *Identification of the specific operation temperatures for the features and development and/or identification of the alloy compositions that meet these needs.* This is a straightforward task for superelastic alloys, as a number of alloy compositions display superelasticity around ambient temperature. However, developing alloy compositions that display shape memory and recovery or two-way shape memory at ambient temperature, and in particular that will activate with simple human input such as touch, will be more difficult.
- *Development of robust, repeatable manufacturing to produce thin-foil TiAl alloys with the necessary stoichiometry control.* As noted, this may entail ingot processing or direct sputtering, both of which currently have some limitations. In particular, ingot processing is subject to variations in stoichiometry across an ingot and difficulty in reducing the ingot size down to the thin-foil dimensions needed for currency applications. Direct sputtering of thin foils is complicated by the difficulty in controlling and reproducing exact alloy stoichiometries, owing either to uneven sputtering rates of Ni and Ti from alloy targets or to uneven spatial distribution from single-component targets, if co-sputtering is used.

A number of university and industrial programs continue to develop NiTi thin-film technology. For example, both General Motors and Johnson Wax currently have active programs.

Estimate of Implemented Production Cost

It is anticipated NiTi security features will be produced by outside vendors and incorporated into currency by the BEP or the substrate manufacturer Crane. As noted previously, wire/foil features might be integrated into notes in the same manner used for the current security threads. Likewise, foil patches could be integrated with notes in the same manner currently used to attach holographic patches on many non-U.S. currencies. This latter approach would require some modification of the BEP production process and would require additional capital investment.

Further Reading

Duerig, T.W., K.N. Melton, D. Stöckel, and C.M. Wayman (eds.). 1990. Engineering Aspects of Shape Memory Alloys. Burlington, Mass.: Butterworth-Heinemann.

Otsuk, K., and X. Ren. 2005. Physical metallurgy of Ti-Ni based shape memory alloys. Progress in Materials Science 50: 511-678.

Zhang, Y., Y.-T. Cheng, and D.S. Grummon. 2005. Indentation stress dependence of the temperature range of microscopic superelastic behavior of nickel-titanium thin films. Applied Physics Letters 98: 033505.

SMART NANOMATERIALS

Description

The field of "smart" nanomaterials may provide the foundation for security features that are simultaneously extremely complex to fabricate and easy to use. Smart nanomaterials are being developed for a wide range of applications. Many of the smart nanomaterials projected to come out of the nanotechnology revolution will be created by some variation of molecular manufacturing (MM). These materials are expected to exhibit a wide range of dynamic behaviors that may be useful for anticounterfeit features.

Many research programs in smart nanomaterials have targeted the use of extremely high technology manufacturing systems to create materials capable of independent dynamic responses. The final output of these responses may be both human-perceptible (detected by simple bioassay) and Boolean—that is, their state would be represented by a simple yes/no answer to the bioassay. If such a smart nanomaterial can be designed in a manner that fits within the physical and fiscal constraints required for currency, the result would be an anticounterfeiting feature almost impossible to counterfeit or simulate, yet as simple to use as color-shifting ink or a watermark.

In the broadest sense, self-assembly describes the natural tendency of physical systems to lose energy to their surroundings and assume patterns or structures of lower energy. Random thermal motions bring constituent particles together in various configurations, so that stable configurations (those with the most binding energy) form, tend to persist, and eventually become predominant. Self-assembly is generally driven by a reduction in the Gibbs free energy for the total system of interest. When properly harnessed, self-assembly offers one of the most reliable, reproducible operating mechanisms known to science and engineering. Many self-assembly events may be viewed as analogous to spontaneous chemical reactions and, as such, the information necessary to produce self-assembly is contained within the fundamental physicochemical structure of the components themselves. Through this simple operation of physical law, a pattern or structure arises in a bounded system with the input of relatively little information from outside. A system slowly approaching equilibrium will assume a simple repetitive structure, while a dynamic system may generate structures of great complexity. For example, molecules in a cooling bucket of water will self-assemble as simple ice crystals, while the same molecules in a turbulent cloud with ever-changing temperature and humidity will self-assemble as complex snowflakes in enormous variety. Many fascinating structures in the natural world around us are self-assembled.

Chemists and biologists often use the term self-assembly in a more restricted sense to describe structure formation in a fluid containing various types of molecules, particularly organic molecules that form weak chemical bonds with a strength that is exquisitely sensitive to molecular shape and orientation. The strongest bond between such molecules often occurs when the molecules fit together in a "lock-and-key" fashion. Biological molecules such as proteins have evolved complex three-dimensional topographies that can create complex higher-order structures based on a "lock-and-key" fit utilizing hundreds or thousands of specific atomic contacts. For example, the bacterial ribosome—a complex molecular machine consisting of about 55 different protein molecules and several ribosomal RNA molecules—will, under appropriate conditions, self-assemble in a test tube. It is important to note that these reactions may occur under conditions that are relatively far from those found in the cytoplasm of living cells. A great deal of nanobiotechnology research is focused on developing ways to stabilize and utilize complex biomolecular self-assembly for processes outside the living cell.

Smart materials whose macroscopic structure depends on the molecular self-assembly of engineered molecular structures may offer a unique combination of "ultrahigh technology" fabrication that produces simple yet uncounterfeitable Boolean behavior. Molecular self-assembly (MSA) is being explored as a manufacturing base for the large-scale industrial processing of consumer items such as computer chips and scaffolds for bioengineering applications like in vitro tissue and organ growth. Large-scale industrial demand for MSA/MM-based facilities may bring the cost security features using the same technology into line with the economic constraints of FRN production the near future. In a sense, these security features could be considered as a spin-off of the National Nanotechnology Initiative.

Feature Motivation

There is a general assumption in the security-device industry that increased technical sophistication results in a decrease in either the simplicity of operation for the end user or the durability of the device itself. For example, the incorporation of electromagnetic security devices requires the use of an instrument capable of reading the signal generated by that device. In the latter case, the incorporation of holograms or other passive optical features increases security via direct visual bioassay, but the materials used to create the holographic image are both subject to physical degradation with repeated handling and may be counterfeited or simulated, since holographic technology is widely available. Smart nanomaterials created by MSA and/or MM may have the ability to combine high-technology, high-durability fabrication methods beyond the capability of most counterfeiters with simple bioassay-based operation.

Nanomaterial-based features are fabricated and/or operate via molecular self-assembly or other dynamic properties that are intrinsic to the composition of the materials themselves. These materials exhibit dynamic behavior, but only in a limited range, triggered by changing a simple physicochemical variable such as thermal energy or the presence (or absence) of a specific chemical or biological compound. The result is posited to be a human-perceptible phenomenon easily and rapidly recognized by the targeted class of users. The usefulness of such a feature is that it recaptures the ease of counterfeit detection inherent in the macroscopic optical security features of FRNs prior to the commercialization of cheap yet sophisticated reprographic products. In addition, because it is the materials themselves that create the feature, the potential exists for bioassays other than those involving vision. This possibility could be useful to the vision-impaired.

Potential Implementations

The Institute for Soldier Nanotechnologies (ISN) at the Massachusetts Institute of Technology (MIT) provides a convenient example of how nanofabrication methods may be used to harness MSA and MM to create technology platforms. There is significant similarity between the performance specifications required for many advanced materials under development for soldier technologies and those required for use as anticounterfeiting features in banknotes. Under field conditions, many of these smart materials will need to be highly durable. In addition, because of the immediacy of the battlefield, many applications (for example: Is there a toxin present? Has a ballistic impact occurred?) will require an almost-instantaneous, unequivocal yes or no, and human-perceptible manifestation. These same traits would make these materials of potential use for anticounterfeit applications.

Example 1: Mechanical Actuators Capable of Switching Between States

The ISN at MIT is "developing nanomaterials that are capable of mechanical actuation and dynamic stiffness."[14] Either or both of these properties could be incorporated into a windowed currency feature that is both visible and tactile. As part of the soldier's battlesuit, these adaptive multifunctional materials will improve soldier survivability. According to the ISN:

> Mechanical actuators embedded as part of a soldier's uniform will allow a transformation from a flexible and compliant material to a non-compliant material that becomes armor, thus protecting the soldier by distributing impact. Soft switchable clothing can also be transformed into a reconfigurable cast that stabilizes an injury such as a broken leg. Contracting materials can be made to apply direct pressure to a wound, function as a tourniquet, or even perform CPR when needed. Mechanical

[14]For further information on this research, see <http://web.mit.edu/ISN/research/team02/index.html>. Accessed February 2007.

> actuators can also be used as exo-muscles for augmentation of a soldier's physical strength or agility and as wound compresses.[15]

Smart, self-assembling materials capable of switching between states may be adaptable as a security feature in future generations of FRNs.

Example 2: Active, Reactive Fiber Coatings

ISN Team 3 is developing smart nanomaterials that will provide protective measures to enable the future soldier to detect and respond to chemical and biological threats.[16] The strategy is to develop

> protective fiber and fabric coatings for integration in the battlesuit. These surfaces will neutralize or significantly decrease bacterial contaminants as well as chemical attack agents such as nerve gas and chemical toxins. For example, some investigations include responsive nanopores that "close" upon detection of a biological agent. In addition, novel organic-inorganic hybrid nanocomposites, consisting of nanoparticles and formed using simple dip processing methods that combine sensing and reactive components.[17]

While the goal of this work is to develop smart nanomaterials that can act as reactive or responsive protective coatings for fibers and fabrics for soldier technologies, any material that can exhibit "dynamic, reversible behavior" that is human-perceptible (that is, may be bioassayed) is obviously a candidate for a security feature in currency.

Materials and Manufacturing Technology Options

These features would use materials and manufacturing methodology that is expected to emerge from recent, large-scale investments in nanoscience and nanotechnology. Importantly, these investments are being driven by the need to nanoscale the manufacturing of components for major industries such as the manufacture of defense-related products, of semiconductor devices for computers and communications, more generally, as well as of bioengineered devices for health care applications.

Also, the ISN is developing the technology to

> enable the synthesis of nanotechnologies developed by ISN to provide soldier protection in the field . . . [using a] broad-based approach to developing processing and fabrication technologies for novel nanomaterials. These technologies must be capable of effectively processing a wide range of components: nanoscale fibers and

[15]See <http://web.mit.edu/ISN/research/team02/index.html>. Accessed February 2007.

[16]For further information on this research, see <http://web.mit.edu/ISN/research/team03/index.html>. Accessed February 2007.

[17]See <http://web.mit.edu/ISN/research/team02/index.html>. Accessed February 2007.

films; multilayered materials; membranes and microdevices; microfluidic devices; functional hollow fibers; and field-responsive materials and devices. Team 5 has as its goal the fabrication and integration of hierarchically structured materials to achieve multiple and synergistic property combinations.[18]

Simulation Strategies

Smart materials, such as those under development at the ISN, will be virtually impossible to counterfeit or simulate without the ability to nanofabricate—that is, to build at the molecular level with atomic precision. The power of a molecular self-assembly-based material or device lies in its simplicity of operation. This simplicity is based on the fact that certain forms of MSA occur spontaneously if and only if one can nanofabricate the materials components of the device ab initio. It is unlikely that this type of technology will become available to any but the most sophisticated government-sponsored counterfeiters for at least a generation.

Nanofabrication facilities necessary to create microprocessor chips or molecular scaffolds for tissue engineering are unlikely to evolve into a format that will make them common household items anytime in the near future. Rather, it is probable that such instrumentation will be controlled by large corporations and major government and/or university research facilities.

Key Development Risks and Issues: Phase I

Smart nanomaterials comprise an emerging area, and a number of possible features can be imagined. The adaptation of the nanomaterials work to counterfeit-deterrent features leverages ongoing programs. Thus, a key risk that would hinder the future development of these novel features would be the termination of the NNI. A second key risk is related to the need to understand any risks associated with environmental, health, and safety effects of nanotechnology. This is an ongoing program at the national level, and it would be beyond the scope of a currency security feature R&D program. The third risk would be an unfocused program that expended resources without achieving concrete results.

Development Plan: Phase I

This feature platform would require a long-term commitment to R&D. The most important activity during Phase I would need to be a comprehensive survey of ongoing work and an analysis of the most-promising targets of opportunity for future counterfeit-deterrent features. Once these targets of opportunity were

[18]See <http://web.mit.edu/ISN/research/team05/index.html>. Accessed February 2007.

selected, a specific plan could be developed for each one. The most desirable situation would be for the BEP to be able to highly leverage ongoing work so that BEP funding would be needed only to adopt the technology, not to invent it.

Estimate of Implemented Production Cost

This technology is too immature to make an estimate regarding production cost. A number of different directions could be pursued. It is possible that this technology could be implemented at some future time for less cost than that of the current security thread. However, usual experience is that new technology costs somewhat more, and a decision must be made about whether the benefit would be worth the added cost.

Further Reading

See the Web site for the Institute for Soldier Nanotechnologies (ISN) at <http://web.mit.edu/ISN/research/> and for the National Nanotechnology Initiative at <www.nano.gov>. Accessed February 2007.

National Research Council. 2006. A Matter of Size—Triennial Review of the National Nanotechnology Initiative. Washington, D.C.: The National Academies Press.

TACTILELY ACTIVE ELECTRONIC FEATURES

Description

A number of material classes offer the potential to develop currency features with tactilely active responses—that is, they will produce local changes in shape or tactile nature as a result of a user input. Two promising classes of material for such features are piezoelectric crystals and electroactive polymers.

A piezoelectric crystal develops a voltage when strained. Conversely, if a voltage is applied to a piezoelectric crystal, it can respond with a strain resulting in a shape change or deflection. This effect can be used in currency applications to produce a user-generated and detectable change in the tactile feel at a specific location on a note.

In the simplest sense, one can envision piezoelectric bumps that would raise or lower on the note when a voltage is applied. The voltage might be supplied by an external source, such as a small battery, but it might also be supplied by an internal source (see the discussion above in the section "e-Substrate").

In practice, while it may be difficult to produce large enough piezoelectric deflections within the constraints of a note, patterns of fine bumps could be produced to generate changes in the tactile nature of the patterned area. The patterning sequence could be varied on notes of different denominations so that each denomination had a unique, readily identifiable pattern.

Since most piezoelectric materials require relatively high voltages for activation, the materials would have to be carefully selected for application in currency. At the present time, it is not clear if tactilely active piezoelectric features can be practically integrated into notes. Such features will require significant development to integrate them effectively into notes with electronic substrates.

Electroactive polymers (EAPs) are currently being developed for application as artificial muscles and in other actuator applications. The phenomena that make these polymers attractive as active materials, that is, they deform under the influence of an applied external voltage, may also have application for active-response features in currency. Upon the application of the external voltage, an EAP will deform over the period of a few seconds.

As an example of the potential of these materials,[19] strips 5 cm × 0.6 cm × 0.02 cm (~0.05 inches thick) can display deflections in the range of 60 to 90 degrees and can displace loads in the range of 50 times the weight of the strips at the ends of the strips. The EAP will regain its original shape upon the reversing of the polarity of the voltage. This process can be repeated over a large number of cycles with

---

[19]For more information, see <http://www.polysep.ucla.edu/Research%20Advances/EAP/electroactive_polymers_as_artifi.htm>. Accessed February 2007.

no apparent degradation in response. The voltages necessary to drive the response are quite low, on the range of a few volts, in the range offered by dry cell batteries (D, AA, AAA, C).

In a currency application, one might envision a number of types of active-response features developed from EAPs. These could be in the form of an active bending security strip or strips or, with more development, features that changed shape or topography (including changes in tactile nature). The note response would be designed to be unique to the denomination of the note. Clearly, the response would require a voltage input that could either be self-contained on the note or provided by the user in the form of a readily available battery. Both of these power options would require integration with some level of an electronic substrate (see the section above entitled "e-Substrates").

Feature Motivation

The active nature of features based on piezoelectric crystals and electroactive polymers would be a strong deterrent to currency counterfeiting at all levels of counterfeiting. Primitive, opportunist, and petty criminal counterfeiters would not be able to simulate the active response of the features accurately. Furthermore, feature configurations and designs specific to each denomination would render of dubious value attempts to "raise" notes through bleaching and reprinting higher denominations. Professional criminal and state-sponsored counterfeiters would be challenged to duplicate the proper materials and processing to obtain the required response. Furthermore, integration with the necessary electronic substrate would be extremely difficult to duplicate.

Active-response features would have a number of advantages for use by the general public. The novel nature of the features would attract attention from the public, leading to people's making use of the features. Currency users could readily detect shape changes and changes in the tactile nature of the features. Denomination-specific placement and patterning of the features would allow simple identification of the note values, making such features very useful for visually impaired users.

Potential Implementations

As described above, tactilely active features would allow point-of-sale user verification of a note's authenticity and denomination by facilitating a user-input/feature-output response. The user would primarily use the change in shape and tactile nature to evaluate the note. However, other human-perceptible indicators such as change in visual reflectivity might also be detectable.

In the simplest form, EAP strips could be integrated into a banknote. Such strips would change shape when the note was folded across the contacts of an AA or AAA battery (both provide 1.5 volts and are pervasively available). Different denominations would have both the thread and contacts in different locations on the notes, leading to very specific and publicly identifiable denominating.

More complex integration into a note would include patterns of either EAPs or piezoelectric crystals that would change local topography on the note with the voltage applied. This patterning could be designed to produce recognizable effects such as buildings, eagles, or numbers specific to the denomination, as illustrated in Figure D-8. The changes in topography would be recognizable either visually or

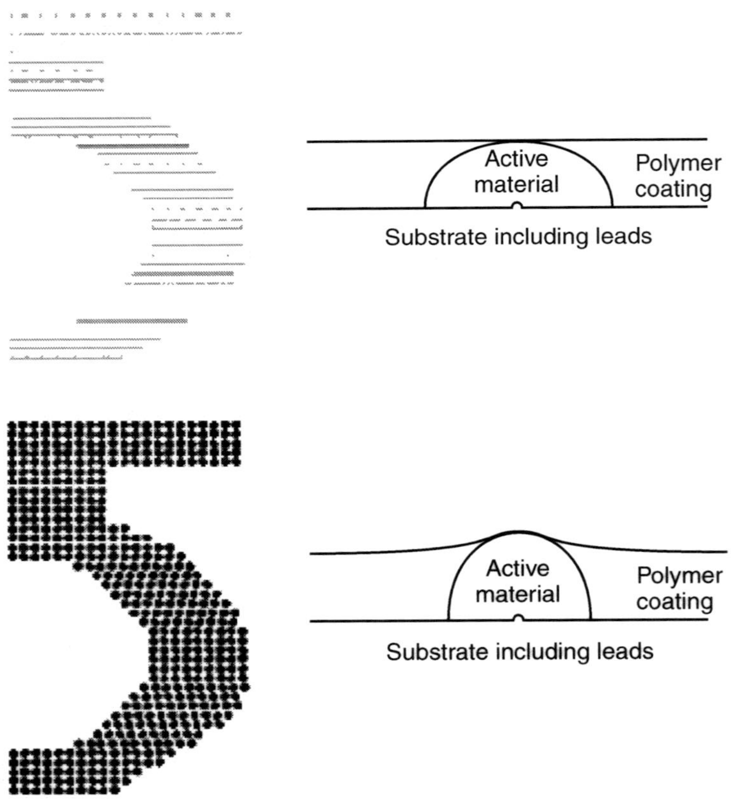

FIGURE D-8 Illustration of the change that would occur in a banknote as the surface relief rose with the application of voltage. In practice the response could be in any direction and still lead to changes in relief.

by changes in their tactile nature. Although the schematic in Figure D-8 illustrates the surface relief rising with the application of voltage, in practice the response could be in any direction and still lead to changes in relief. Thus, the electroactive material would not have to be orientation-controlled during processing. Ideally, the user input could be generated by an on-note power supply.

The changes in tactile nature on the note would be a function of a number of BEP- or supplier-controllable variables, including electroactive material, dot size, and coating. Primitive, opportunist, and petty criminal counterfeiters would not be able to simulate this active change in tactile response accurately. State-sponsored and some professional criminal counterfeiters would be able to reverse-engineer the feature and in time simulate the response.

This feature is primarily targeted to assist currency verification by the general public and point-of-sale merchants. The outstanding benefit of this feature is that it is user-induced (preferably by an integrated power source) and detectable by humans. If the power source can be integrated into the note, the authenticity of the note can be subtly checked without visual inspection, allowing a wide range of users to verify the notes, including the visually impaired.

Materials and Manufacturing Technology Options

Piezoelectric materials are used in a wide range of industrial, consumer, and laboratory applications, and extensive research is currently producing and studying nanoscale piezoelectric materials. For example, piezoelectric print heads are often used as print heads in ink-jet printers; these heads are in fact manufactured through an ink-jet process. Consequently, processing the piezoelectric materials is well established. The key in the current application would be to make the best selection of piezoelectric material to obtain the intended response and to properly integrate the patterned feature with the electronic substrate.

A wide range of electroactive polymers is being developed, including ferroelectric polymers, dielectric EAPs, electrostrictive graft elastomers, ionic polymer gels, and ionometic polymer-metal composites, all with advantages and disadvantages. Many of these require the polymer to be saturated with water or immersed in water. Clearly, the need for water to achieve the mechanical response has implications for currency applications. Further advances in EAP technology or flexible encapsulation will be necessary for some EAPs to be integrated into currency. Thus, this is a relatively immature technology that will require considerable development.

Simulation Strategies

While piezoelectric disks are available for less than $1 at hobby stores, engineered patterns of piezoelectric features would be difficult for many counterfeiters to simulate. The proper patterning and integration with the appropriate leads and power supplies would require processing beyond the realm of all but professional criminal or state-sponsored counterfeiters.

Primitive and opportunist counterfeiters would not be able to duplicate and would have a very difficult time simulating the active response of EAPs. Simulations would most likely be based on using elastic polymers to simulate some aspects of the response. However, it is unlikely that such simulations would be reversible, and they would not hold up under even moderate scrutiny. These efforts would be further complicated by more complex patterns that would result in topography changes in the notes. Petty and professional criminals would be challenged in attempting to counterfeit EAPs. The polymer chemistry and processing would require significant technology and equipment to duplicate. It is difficult to envision other approaches to achieving the active response of EAPs that would be any less challenging to the criminal. State-sponsored counterfeiters could, with time, reverse-engineer the technology necessary to obtain the EAPs. However, this would not be a trivial matter. Thus, if proper technology control were in place, state-sponsored counterfeiters would be extremely challenged by active-response features such as EAPs.

Key Development Risks and Issues: Phase I

The integration of electroactive material features would have some impact on the aesthetics of U.S. currency. It is envisioned that such features would be contained in a moderate-sized strip or patch, approximately the size of the current security strip or of a fingertip. The feature would be noticeably different from the rest of the note, in the same manner that holographic films, clear windows, or metallic ink are noticeably different from the surrounding areas on current currencies. Furthermore, it is envisioned that both internal or external power sources would require recognizable features that would impact the aesthetics of the note. Although aesthetics is a subjective issue, it is likely that such features would be judged to degrade the aesthetics of the currency.

Electroactive features would be socially neutral, although any incorrect suggestion that the feature might actually record or track the fingerprints or DNA, or add what you would like, of anyone who touched the note could affect its social acceptability. Many products have been doomed by false rumors about them.

Most electroactive polymers are in the early stages of development and are a considerable way from being used as artificial muscles. However, other EAPs are at or near implementation in other actuator technologies. In order for EAPs to be integrated into currency, a number of issues would need to be addressed. Viable EAPs that would be useable in a range of environments, from dry to damp, would need to be developed—possibly requiring new types of EAPs or the encapsulation of the EAPs. The durability of EAPS should be quite high, given the thin sections and large deflections currently reported. However, this might be compromised to a degree by encapsulation.

The durability of tactilely active piezoelectric features is unknown and would have to be assessed as part of the development of such features. Because such features would be electronic in nature, durable electronic platforms would be necessary, in addition to durable active features. It is envisioned that considerable redundancy of leads and fine-scale piezoelectric bumps would be necessary.

Many of the key technological challenges to developing electroactive currency features go hand in hand with those of developing electronic substrates. For example, currently piezoelectric patterns can be produced in manufacturing settings. The difficulty would be to produce patterns integrated with the electronic substrate. Furthermore, the need to power the devices effectively, either through external sources, or preferably, through integrated power sources, would take significant development in order to integrate either EAPs or piezoelectric crystals into currency.

Development Plan: Phase I

All of the functions critical to implementing electroactive features into currency—piezoelectric devices, EAPs, the electronic substrate, and power supplies—have been demonstrated individually. However, integration of the technologies has not been developed in a currency platform.

There are significant industrial and university programs developing e-substrate technologies that would be well suited to developing electroactive features for currency. Also, at the present time some work is going on to develop a number of different variable-friction surface technologies. However, these do not appear to be targeted to currency applications.

A key milestone in the development of this technology would be the patterning and encapsulation or coating of the piezoelectric or EAP features in a durable form suitable for integration in currency. At that point the effectiveness of the active changes in shape and tactile nature could be assessed.

Estimate of Implemented Production Cost

This technology is very immature for currency applications, and thus an analysis of the implementation cost should be included as part of the development program during Phase I. However, it would be expected that features based on this technology will be significantly more costly compared with the current security thread. Since the devices would be embedded in the substrate, the substrate manufacturer would have to make significant investments in process technology.

The impact on manufacturing operations within the BEP might be minimal, but it is premature to state this unequivocally.

Further Reading

Bar-Cohen, Y. 2001. ElectroActive Polymers—EAPs. Available at <http://www.azom.com/details.asp?ArticleID=885#_Ferroelectric_Polymers>. Accessed February 2007.

Bar-Cohen, Y. 2006. WorldWide Electroactive Polymer Actuators Webhub. Available at <http://eap.jpl.nasa.gov>. Accessed February 2007.

Cohen, J.Y. Electroactive Polymers as Artificial Muscles—A Primer. Available at <http://www.polysep.ucla.edu/Research%20Advances/EAP/electroactive_polymers_as_artifi.htm>. Accessed February 2007.

Free electricity from nanogenerators. MIT Technology Insider. Available at <http://www.techreview.com/read_article.aspx?id=16746&ch=nanotech>. Accessed February 2007.

Galassi, C., M. Dinescu, K. Uchino, and M. Sayer (eds.). 2000. Piezoelectric Materials: Advances in Science, Technology and Applications. NATO Science Partnership Sub-Series: 3: High Technology, Vol. 79. Norwell, Mass.: Kluwer Academic Publishers.

Price, B., and C. Blankenship. 2003. NASA Langley Research Center is offering to license its intellectual property in electroactive polymers to prospective industrial clients for commercialization opportunities. Available at <http://www.teccenter.org/electroactive_polymers/index.html>. Accessed February 2007.

Wang, Z.L., and J. Song. 2006. Piezoelectric nanogenerators based on zinc oxide nanowire arrays. Science 312(April 14): 242-246.

E

An Example Using a Flow Model for Counterfeiting

This appendix contains an example of the kind of analysis suggested by the flow model of the currency circulation system discussed in Chapter 2 and shown in Figure 2-2, duplicated as Figure E-1 in this appendix. The variables used to represent this system are defined in Table 2-5, duplicated as Table E-1 in this appendix.

Table E-2 presents the flow equations for each repository of counterfeit currency. This example is based on a number of simplifying assumptions.

- The analysis holds for a given combination of (class of counterfeiter) × (note denomination) × (country of passing) × (etc.).
- There is no interaction among classes, denominations, and so on, so that combined effects can be obtained by adding.
- The parameters of the model are essentially constant over time.
- The fraction of successful pass attempts s (and the related removal fractions, r and h) is the same for initial and subsequent pass attempts. (In reality, of course, the first pass attempt may differ from repass attempts in a number of ways. Arguments can be made that the first pass attempt is *easier* because of the perpetrator's careful selection of the venue, or *harder* owing to more suspicious circumstances.)
- Culling in the banking system—presumably by means of machine authentication—is perfect. This assumption has an important implication: Even though a feature may make culling easier, if it does not decrease the passing parameter s, the feature will have no effect on the amount of counterfeit currency in circulation.

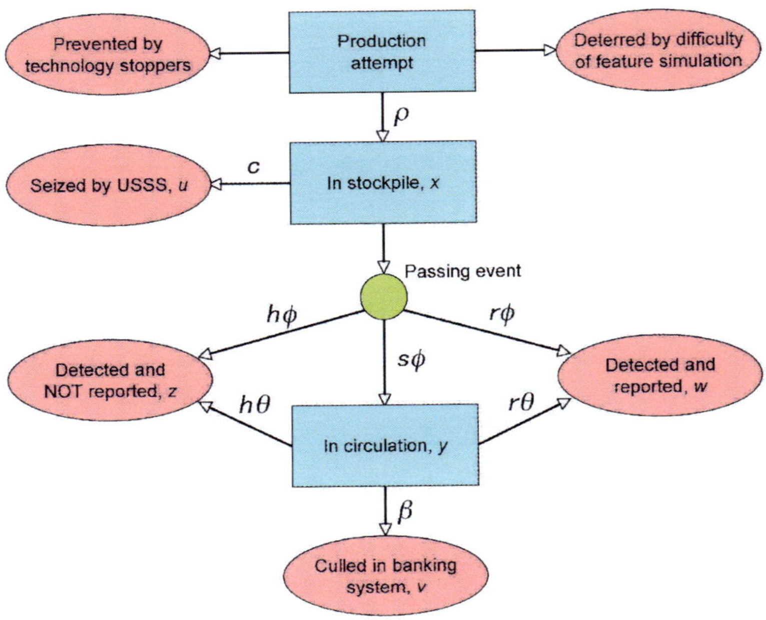

FIGURE E-1 A flow model for counterfeiting. The boxes indicate the components of the counterfeiting system. The ovals represent processes by which counterfeit notes are removed from the system. Arrows indicate the flow of counterfeit notes. NOTE: USSS = U.S. Secret Service. Variables are defined in Table E-1.

TABLE E-1 Variables in the Flow Model for Counterfeiting

Variables	Definition
Counterfeit repository (total $)	
u	$ seized from stockpile by U.S. Secret Service (USSS)
v	$ culled from banking system by the Federal Reserve System (FRS)
w	$ turned in to USSS by a recipient
x	$ in stockpile, awaiting first pass attempt
y	$ in circulation
z	$ held by a recipient, but not reported or repassed
Rates	
c	Confiscation rate of notes from stockpile ($ per unit time)
β	Rate of culling in the banking system by the FRS (per unit time)
ϕ	Rate at which first attempts are made to pass note from stockpile (per unit time)
θ	Rate at which repass attempts are made during circulation (per unit time)
ρ	Counterfeit production rate ($ per unit time)
Other parameters	
h	Fraction of pass attempts held and not reported to the USSS; $h = 1 - s - r$
r	Fraction of pass attempts reported by recipient
s	Fraction of successful pass attempts
t	Time since start of production

TABLE E-2 Differential, Time-Dependent, and Steady-State Flow Equations in the Flow Model for Counterfeiting

Repository	Flow Rate In/Out of Repository	Time-Dependent Amount in Repository	Steady-State Amount in Repository
x	$\frac{dx}{dt}=\rho-c-\phi x$	$x(t)=\frac{\rho-c}{\phi}[1-\exp(-\phi t)]$	$x(\infty)=\frac{\rho-c}{\phi}$
y	$\frac{dy}{dt}=s\phi x-(h+r)\theta y-\beta y$	$y(t)=\frac{s(\rho-c)}{\phi-\mu}\left[\exp(-\phi t)-\exp(-\mu t)+\frac{\phi-\mu}{\mu}\right]$	$y(\infty)=\frac{s(\rho-c)}{\mu}$
z	$\frac{dz}{dt}=\phi hx+\theta hy$	$z(t)=h(\rho-c)\left[\frac{\exp(-\phi t)}{\phi}+t\right]+\frac{\theta hs(\rho-c)}{\phi-\mu}\left[\frac{\exp(-\mu t)}{\mu}-\frac{\exp(-\phi t)}{\phi}+\frac{\phi-\mu}{\mu}t\right]$	
w	$\frac{dw}{dt}=\phi rx+\theta ry$	$w(t)=r(\rho-c)\left[\frac{\exp(-\phi t)}{\phi}+t\right]+\frac{\theta rs(\rho-c)}{\phi-\mu}\left[\frac{\exp(-\mu t)}{\mu}-\frac{\exp(-\phi t)}{\phi}+\frac{\phi-\mu}{\mu}t\right]$	
u	$\frac{du}{dt}=c$	$u(t)=ct$	
v	$\frac{dv}{dt}=\beta y$	$v(t)=\frac{\beta s(\rho-c)}{\phi-\mu}\left[\frac{\exp(-\mu t)}{\mu}-\frac{\exp(-\phi t)}{\phi}+\frac{\phi-\mu}{\mu}t\right]$	

NOTE: Variables are defined in Table E-1. Note that for convenience, $\mu \equiv (h + r)\theta + \beta$.

With these assumptions, the flows can be represented by first-order, linear ordinary differential equations that are straightforward to solve. Table E-2 also shows the time-dependent and steady-state (that is, $t \to \infty$) solution for each repository. The steady-state solutions give the "average" amount of counterfeit currency in each of the counterfeit repositories. (Note that the steady-state solution for the removal repositories is necessarily infinite, since counterfeits only flow in and never out.) Of greatest interest are $x(\infty)$, the average amount of counterfeit in the stockpile waiting to be passed, and $y(\infty)$, the average amount of counterfeit in circulation. These are the amounts that arguably would be decreased by using effective features.

FEATURE ANALYSIS USING THE FLOW MODEL

For the sake of demonstrating the power of the flow model proposed, consider three potential features with different characteristics, as shown in Table E-3. For example, Feature A might use a metameric ink that can be easily simulated by a laser printing process but easily detected with a handheld device. Feature B might be similar to current security threads. Feature C might be a form of extreme microprinting, which is hard to duplicate but also hard to detect in a casual transaction.

A naïve comparison of these features might assign values on a scale of 0 to 2 and add values to determine an "effectiveness score" (see Table E-4). That the total score is the same for all three features points out the inherent dangers in using additive measures—something that might not be apparent with a larger, richer, and more complex set of criteria.

By contrast, it is possible to examine the three features using Equation (2-1) in Chapter 2. Assume, for convenience, that $\theta = 50$ (that is, the average time between passing attempts is roughly 1 week), and that $\beta = 10$ (the average time to get swept up by a Federal Reserve Bank is roughly 1 month). In addition, assume that the stockpile confiscation rate c is proportional to the production rate ρ, so that $c = k\rho$, where k is a constant of proportionality. Then Equation (2-1) becomes

$$y(\infty) = \frac{(1-k)s\rho}{(1-s)50+10} \propto \frac{s\rho}{(1-s)5+1} \qquad \text{(E-1)}$$

TABLE E-3 Potential Features for Flow Model

Feature	Difficulty of Simulation	Detection Effectiveness
A	Low	High
B	Medium	Medium
C	High	Low

TABLE E-4 Demonstration of a Possible Assignment of Effectiveness Scores/Values

Feature	Difficulty of Simulation	Detection Effectiveness	Total Effectiveness
A	0	2	2
B	1	1	2
C	2	0	2

TABLE E-5 A Nominal Characterization Table for the Comparison of Features A, B, and C

Feature	Production Rate ρ	Passing Fraction s	Amount Circulating $y(\infty)$
A	0.9	0.1	0.02
B	0.5	0.5	0.07
C	0.1	0.9	0.06

Comparison tables can now be developed to reflect the respective features' effects on the remaining parameters s and ρ.[1]

To continue the hypothetical example, suppose a "nominal characterization" table could be used to assign values for s and ρ, as shown in Table E-5 (where the values of ρ are scaled so that $\rho = 1$ reflects the lowest difficulty to simulate, and $\rho = 0$ the highest).

Note that while the additive score for the parameters that measure difficulty of simulation (ρ) and detection efficiency (s) remains the same for all features, the steady-state amounts of counterfeits in circulation differ by more than a factor of three. According to this metric, Feature A is much more attractive than the other two: in this example, hindering passing is more important than deterring production in keeping counterfeits out of circulation. Interestingly, Feature B, which nominally deters both production and passing moderately well, gives the highest amount of counterfeits in circulation: at least in this example, it is more important to do one thing well than to do all things adequately.

One might ask, What is the effect of a note that does both things well? For a note design that deters production ($\rho = 0.1$) and hinders passing ($s = 0.1$), Equation (E-1) predicts a circulating amount $y(\infty) = 0.002$, a factor of 10 less than for Feature A, which only does one thing well. Unsurprisingly, the best feature is one that deters both production and passing; therefore, in evaluating proposed features, the committee selected ones that scored well in both areas, as discussed in Chapter 4.

The flow model feature analysis encourages other investigations. For example, each class of counterfeiter will likely have different values of each variable (that is, both the production rate and pass frequency are likely higher for the professional than for the opportunist), and this affects where and how deterrence and enforcement measures are optimally applied. Thus, the flow model not only can provide a quantitative basis for deciding on and supporting features to develop, but it can serve as a means for directing research into the values of critical parameter values.

[1]Doing this in a realistic and supportable fashion requires experimental and experiential data that are not available to this committee; the Bureau of Engraving and Printing could undertake a research program to obtain these data.

In particular, it would seem appropriate for the Bureau of Engraving and Printing (BEP) to engage in a series of large-scale field-like experiments in order to understand the "detectability" of various existing and proposed features.

The committee also observes that the model makes it obvious that machine authentication affects counterfeit flows only via culling in the banking system (given by the variable β). While counterfeit notes may pass through (and be rejected by) machine readers a number of times during circulation, these are almost always returned to the holder and remain in circulation until removed by a human or culled in the banking system. Therefore, machine readers only weakly affect r or h.

Finally, it should be clear that the hypothetical example presented in this appendix should not be interpreted to mean that the committee is suggesting that a single, dominant feature by itself would deter counterfeiting. The whole purpose of presenting a modeling approach is to provide a framework within which the effectiveness of combinations of features can be assessed in a consistent manner.

F

Biographical Sketches of Committee Members

Robert E. Schafrik, *Chair,* is currently the general manager of the Materials and Process Engineering Department at GE Aviation. He is responsible for the development of advanced materials and processes used in GE's aeronautical turbine engines and their marine and industrial derivatives. He oversees materials application and engineering activities supporting design engineering, manufacturing, and field-support activities worldwide. He also operates a state-of-the-art laboratory for advanced materials development, characterization, and failure analysis. Before joining GE in November 1997, Dr. Schafrik served in two concurrent positions within the National Research Council, which he joined in 1991: director of the National Materials Advisory Board and director of the Board on Manufacturing and Engineering Design. He also served in the U.S. Air Force. His career highlights there included research metallurgist, manufacturing technologist, materials application engineer, manager of F-16 engine programs, and headquarters manager of air superiority weapons programs. Dr. Schafrik has a Ph.D. in metallurgical engineering from Ohio State University, an M.S. in information systems from George Mason University, an M.S. in aerospace engineering from the Air Force Institute of Technology, and a B.S. in metallurgy from Case Western Reserve University.

Martin A. Crimp is a professor of chemical engineering and materials science at Michigan State University. He is an expert in the development and use of a variety of advanced characterization and imaging tools. His research applies a variety of innovative analysis tools to describe and illustrate the design, performance, and failure of advanced materials. Some examples include the creative use of scanning electron microscopy to image atomic-scale features through electron channeling contrast imaging; the use of electron energy-loss spectroscopy to study environmental and radiation effects on carbon nanotubes; and the use of advanced transmission electron microscopy techniques such as convergent-beam electron diffraction to study the growth of nanowires and nanostructures. Dr. Crimp has a Ph.D. from Case Western Reserve University and M.S. and B.S. degrees from Michigan Technological University.

Charles B. Duke is professor of physics at the University of Rochester. At the end of 2005 he retired as vice president and senior research fellow in the Xerox Innovation Group. Prior to holding that position, he was deputy director and chief scientist of the Pacific Northwest Division of the Battelle Memorial Institute and affiliate professor of physics at the University of Washington. From 1972 to 1988 he held various technical and management positions at the Xerox Research Laboratories in Webster, New York, and was an adjunct professor of physics at the University of Rochester. During the years 1969-1972, he was a professor of physics and member of the Materials Research Laboratory and Coordinated Science Laboratory at the University of Illinois in Urbana-Champaign, following 6 years as a staff member of the General Electric Corporate Research and Development Center in Schenectady, New York. He received his Ph.D. in physics from Princeton University in 1963, following a B.S. summa cum laude with distinction in mathematics from Duke University in 1959. He is a fellow and an honorary member of the American Vacuum Society, a fellow of the American Physical Society, a fellow of the Institute of Electrical and Electronics Engineers (IEEE), a member of the Materials Research Society, and a life member of Sigma Xi. In 1977, Dr. Duke received the Medard W. Welch Award in Vacuum Science and Technology. He served as president of the American Vacuum Society in 1979, on its board of directors for 7 years, and as a trustee during 2003-2005. In 1981 he was named one of the ISI 1000 internationally most cited scientists. During the period 1985-1986 he served as founding editor-in-chief of the *Journal of Materials Research,* and from 1992 to 2001 he was editor-in-chief of *Surface Science* and *Surface Science Letters.* He served on the council of the Materials Research Society for 7 years, serving as treasurer in 1991-1992. In 1993 he was elected to the National Academy of Engineering and in 2001 to the National Academy of Sciences. During the period 1995 to 1999 he served on the council and executive board of the American Physical Society. In 2006 he was awarded the George E. Pake prize of the American Physical Society. From 1997 to 2000 he served as general chairman of the Physical Electronics Conference. He served on the governing board of the American Institute of Physics for 11 years and on its corporate associates advisory committee for nearly 20 years. During 2004-2005 he served as chair of a National Research Council study of Network Science. He has written more than 370 papers on surface science, materials research, semiconductor physics, and the electronic structure of molecular solids; several patents on the use of feedback in the design of digital imaging and printing systems; and he has written a monograph on electron tunneling in solids and has edited three books: *Surface Science: The First Thirty Years* (1994), *Color Systems Integration* (1998), and *Frontiers in Surface and Interface Science* (2002).

Alan H. Goldstein is a professor of biomaterials and holds the Fierer Chair of Molecular Cell Biology at Alfred University. His work focuses on structure-function studies of protein binding to materials surfaces, including both biochemical and molecular modeling approaches. Dr. Goldstein has published extensively on the topic of converging technologies in biology, nanotechnology, information technology, and cognitive sciences, and his research ranges from protein engineering to biomimetic materials to mineral phosphates. Specific projects include the biochemistry of interactions between glass fibers and the extracellular matrix; molecular mechanics and molecular dynamics modeling to simulate the adsorption to materials surfaces; and the application of electrophoretic methods to study protein adsorption layers on glass and ceramic surfaces. Dr. Goldstein holds a B.Sc. from New Mexico State University in agronomy and a Ph.D. in genetics from the University of Arizona. He received the Biotechnology Faculty Research Award in 1995 from the California State University Program for Education and Research in Biotechnology.

Elizabeth A. Holm is a distinguished member of the technical staff in the Computational Materials Science and Engineering Department at Sandia National Laboratories. She is a computational materials scientist with a longstanding interest in bringing materials modeling to industrial practice. Over her 14 years at Sandia, she has worked on simulations to improve processes to make materials for advanced lighting, on prediction of microcircuit aging and reliability, and on the processing of innovative bearing steels. Her research areas include the theory and modeling of microstructural evolution in complex polycrystals, the physical and mechanical response of microstructures, and the wetting and spreading of liquid metals. Dr. Holm obtained her B.S.E. in materials science and engineering from the University of Michigan, an S.M. in ceramics from the Massachusetts Institute of Technology, and a dual Ph.D. in materials science and engineering and scientific computing from the University of Michigan. She has received several professional honors and awards, is a fellow of ASM International, and serves on the National Materials Advisory Board and the board of directors of the Minerals, Metals, and Materials Society. Dr. Holm has authored or coauthored more than 90 publications.

Pradeep K. Khosla is currently the dean of the College of Engineering and the Philip and Marsha Dowd Professor in the College of Engineering and School of

Computer Science at Carnegie Mellon University. Dr. Khosla's research interests are in the areas of real-time embedded systems, secure embedded software, biometrics, and intelligent systems. He is the recipient of several awards and honors, including the McDowell Award from the IEEE Computer Society, the ASEE Westinghouse Award for Education in 1999, and the Siliconindia Leadership Award. Dr. Khosla is a fellow of IEEE, the American Association for the Advancement of Science, and the American Association for Artificial Intelligence. Dr. Khosla is a member of the National Academy of Engineering. He is a consultant to several companies and venture capitalists, has served on the technology advisory boards of many start-up companies, and currently serves on several advisory boards, including iNetworks, ITU Ventures, iPolicy, and Alcoa Chief Information Officer's Advisory Board. He has served as a member of the Strategy Review Board for the Ministry of Science and Technology, Taiwan; the information technology advisory committee, Australia's Commonwealth Scientific and Industrial Research Organisation; the Council of Deans of the Aeronautics Advisory Committee, NASA; and Senior Advisory Group, DARPA Program on Joint Unmanned Combat Air Systems. Dr. Khosla has coauthored two books and is the author or coauthor of more than 300 publications. He received a B.Tech. from the Indian Institute of Technology in Kharagpur, India, and M.S. and Ph.D. degrees from Carnegie Mellon University.

Carolyn R. Mercer is an aerospace engineer at the NASA Glenn Research Center. Her research interests include the development of optical instrumentation to measure fluid and surface behaviors. She is currently responsible for developing technologies for integrated vehicle health management and has managed projects in adaptive propulsion systems, intelligent propulsion controls, and instrumentation systems for aerospace ground testing. Dr. Mercer's specific experience includes using optics to measure the flow inside engines to test designs for improving fuel economy, using structured laser illumination to measure the shape of solid surfaces for manufacturing processes, and devising a liquid-crystal device to measure fluid temperature, density, or concentration for microgravity science. Dr. Mercer has a bachelor's degree in aeronautic and astronautic engineering from Ohio State University, an M.S. degree in physics from Cleveland State University, and a Ph.D. in optical science from the University of Arizona. She has published more than 30 papers, holds two patents, and has edited a book entitled *Optical Metrology for Fluids Combustion and Solids* (2003).

Stephen M. Pollock is an emeritus professor at the University of Michigan and an international scholar in the mathematical modeling of systems, sequential decision analysis, and operations research. His work in understanding how to make critical trade-offs in complex decision-making processes has been applied to such diverse problems as military search and detection, manufacturing process monitoring, and design of adaptive radiation treatment plans. He has a B.Eng.Phys. from Cornell University and an M.S. in physics and Ph.D. in physics and operations research from the Massachusetts Institute of Technology. Dr. Pollock has been involved in teaching and applying a wide variety of operations research methods, with the aim of understanding and influencing operational phenomena in industrial and military settings, as well as in the public sector, medicine, and biology. He has authored more than 60 technical papers, has coedited two books, and has served as a consultant to more than 30 industrial, governmental, and service organizations. He has been associate editor and area editor of *Operations Research*, senior editor of *IIE Transactions*, associate editor of *Management Science*, and on the editorial boards of other journals. He has served on advisory boards for the National Science Foundation, as a member of the Committee on Applied and Theoretical Statistics of the National Research Council (NRC), and as a member of the Army Science Board. He was president of the Operations Research Society of America in 1986, was awarded the 2001 Kimball Medal by the Institute for Operations Research and the Management Sciences (INFORMS), and was named a founding INFORMS Fellow in 2002. Dr. Pollock is a member of the National Academy of Engineering.

Arthur J. Ragauskas is a fellow of the International Academy of Wood Science and the Technical Association of the Pulp and Paper Industry (TAPPI). His research program at the Georgia Institute of Technology is directed at understanding and exploiting innovative sustainable lignocellulosic materials. This multifaceted program seeks to develop new and improved applications for nature's premiere renewable biopolymers, including cellulose, hemicellulose, and lignin. Dr. Ragauskas has been a Luso-American Foundation teaching fellow at the Universidade da Beira Interior, Portugal; an invited guest teaching professor at Chalmers University of Technology, Sweden, and the South China University of Technology. He has authored more than 185 papers, patents, and conference proceedings. He is an associate editor for the *Journal of Pulp and Paper Science, Holzforschung, Journal of Wood Chemistry and Technology*, and has served on several advisory boards and review panels including those for the European Commission Research Directorate, J. Paul Getty Trust, TAPPI, the National Science Foundation, and the U.S. Department of Agriculture and the U.S. Department of Energy. Dr. Ragauskas obtained his honors B.Sc. degree in chemistry in 1980 and his Ph.D. in 1985 from the University of Western Ontario.

John A. Rogers is Founder Professor of Engineering at the University of Illinois at Urbana-Champaign, with appointments in the Departments of Materials Science and Engineering, Electrical and Computer Engineering, Chemistry, and Mechanical Science and Engineering. His research interests are in the field of unconventional material patterning techniques for photonics and electronics. This work combines fundamental studies with forward-looking engineering efforts in a way that promotes positive feedback between the two. Some highlights of his work include the first flexible displays and gigahertz electronics on plastic, tunable microfluidic optical fiber, stretchable forms of single-crystal silicon, and techniques for large-area, three-dimensional nanofabrication. Dr. Rogers received degrees in physics and in chemistry from the University of Texas at Austin and a Ph.D. degree in physical chemistry from the Massachusetts Institute of Technology (MIT). In addition to more than 150 publications, he has nearly 60 patents and patent applications in areas ranging from acoustics to neural networks to nanofabrication to fiber optics and organic electronics. More than 30 of these are licensed or in active use. Dr. Rogers has received many awards for his research, including, most recently, recognition from MIT's *Technology Review* magazine as inventor of one of the top 10 most significant technologies for 2005 (stretchable silicon) and from *Scientific American* as one of the top 50 research leaders for 2005.

Barton Rubenstein creates indoor and outdoor sculpture with and without water for public and private spaces, including corporate, commercial, and academic institutions as well as private residences. He typically works with bronze, stainless steel, stone, and glass. Dr. Rubenstein has received several awards for his artwork, lectures frequently about his work, and has been featured in numerous newspaper articles across the country. He has worked with various art forms throughout his career, including lithography, etching, woodcuts, architectural drawing, and sculpture. He trained in physics and mechanical engineering at Haverford College, Pennsylvania, and then completed his M.Sc. and Ph.D. degrees at the Weizmann Institute of Science, Israel, studying the brain and visual sciences. This research in neuroscience focused on how people visually perceive the world. His research attempted to elucidate various anomalies of visual perception, such as camouflage and, more generally, the processes at work within the visual system. Published work of his research has appeared in *Science*, *Journal of the Optical Society of America*, and *Scientific American*, as well as in *Time Magazine* and on National Public Radio. He is the principal of Rubenstein Studios.

Michael A. Smith is the director of research and university alliances at France Telecom R&D, San Francisco. He is a specialist in video content analysis and the author of numerous papers and a book on the subject. His research interests include visualization and indexing for multimedia libraries; multimodal audio and video processing; media interfaces between people and machines for mobile and fixed platforms; and e-learning for disadvantaged communities. His innovations include patented video analysis and summarization technology, which is licensed by media management companies. Before joining France Telecom, Dr. Smith founded AVA Media Systems and worked as a visiting professor in the Computer Vision Research

Center at the University of Texas at Austin. He has worked as a visiting professor at Morehouse College in Atlanta and at the University of Campinas in Brazil and as a guest lecturer at the University of California at Berkeley. Dr. Smith is on several advisory boards that support developing nations such as Trinidad and South Africa through digital media technology. He is also the director for an NSF program on broadening participation by underrepresented students pursuing graduate degrees in computer science. Dr. Smith holds a Ph.D. in electrical and computer engineering from Carnegie Mellon University, an M.S. in electrical engineering from Stanford University, and a B.S. degree in electrical engineering from North Carolina A&T University and Tuskegee University.

Gary K. Starkweather received his B.S. in physics from Michigan State University in 1960 and a master's degree in optics from the University of Rochester in 1966. He has spent more than 40 years in the imaging sciences and holds more than 44 patents in the fields of imaging, color and hard-copy devices. From 1962 to 1964, he worked for Bausch & Lomb, Inc., in Rochester, New York. From 1964 until 1988 he was employed by Xerox Corporation, where he became a senior research fellow. While at the Xerox Palo Alto Research Center (PARC), he invented the laser printer. Dr. Starkweather has received a number of awards for this work, including the Xerox President's Achievement Award (1977), the Johann Gutenberg Prize from the Society for Information Display (1987), and the David Richardson Medal from the Optical Society of America (1991). From 1988 until 1997, he was employed by Apple Computer as an Apple Fellow involved in publishing and color imaging products and research. In 1994, he received a Technology Academy Award for his consulting work with Lucasfilm and Pixar on color film scanning. In 2002, he was inducted into the Technology Hall of Fame at COMDEX. He has recently retired from Microsoft research as an architect working on displays and information processing. He has published many papers and has written a book chapter entitled "High Speed Laser Printers" for Academic Press. He continues to serve on several technical committees involved in display and color-related imaging issues and has lectured at both Stanford University and the University of California at Los Angeles. He is a member of the National Academy of Engineering.

Dennis J. Trevor is technical manager of the Optical Materials Group of OFS Laboratories in Murray Hill, New Jersey, formerly part of Lucent Technologies Bell Laboratories. He led the development and assisted in the implementation of the first commercial Sol-Gel process used in optical-fiber preform manufacture. Currently he is developing new applications using this technology in related fields of photonics. His additional work in material growth methods includes chemical reaction studies of metal clusters, active oxygen source growth of high-T_c superconductor films, and very low pressure chemical vapor deposition of silicon and germanium alloys. In his 15 years at Bell Laboratories, Dr. Trevor has also developed experimental methods to improve our understanding in a wide range of fields, from semiconductor plasma processing to surface diffusion in electrochemical corrosion. He received his B.S. in chemistry at the Illinois Institute of Technology in 1975 and his Ph.D. in physical chemistry from the University of California at Berkeley in 1980. He has more than 50 publications and holds several patents.